THE STUDENT'S GUIDE TO SYSTEMATIC BOTANY

THE STUDENT'S GUIDE TO SYSTEMATIC BOTANY

INCLUDING THE CLASSIFICATION OF PLANTS AND DESCRIPTIVE BOTANY

ROBERT BENTLEY

GYAN

Published by

Gyan Publishing House
5, Ansari Road
Daryaganj, New Delhi-110002
Phone: 011-47034999, 9811692060
E-mail: books@gyanbooks.com

Distribution Network
gyanbooks.com
India, USA, Canada, UK, Australia

ISBN: 978-93-7414-050-5 (PB)

First Published, 1884

2nd Impression 2026

Printed at: Gyan Press, Delhi.

THE STUDENT'S GUIDE TO SYSTEMATIC BOTANY

Author: ROBERT BENTLEY

THE STUDENT'S GUIDE

TO

SYSTEMATIC BOTANY

By the same Author.

A MANUAL OF BOTANY; including the Structure, Functions, Classification, Properties, and Uses of Plants. Fourth Edition, with 1,185 Engravings. Crown 8vo. 15*s*.

THE STUDENT'S GUIDE TO STRUCTURAL, MORPHOLOGICAL, AND PHYSIOLOGICAL BOTANY. With 660 Engravings. Fcp. 8vo. 7*s*. 6*d*.

And, jointly with Dr. Trimen.

MEDICINAL PLANTS: being descriptions, with original Figures, of the Principal Plants employed in Medicine, and an Account of their Properties and Uses. In 4 vols. large 8vo. with 306 Coloured Plates, bound in half-morocco, gilt edges, £11. 11*s*.

THE STUDENT'S GUIDE

TO

SYSTEMATIC BOTANY

INCLUDING THE

CLASSIFICATION OF PLANTS AND DESCRIPTIVE BOTANY

BY

ROBERT BENTLEY, F.L.S., M.R.C.S. Eng.

FELLOW OF KING'S COLLEGE, LONDON
HONORARY MEMBER OF THE PHARMACEUTICAL SOCIETY OF GREAT BRITAIN
HONORARY MEMBER OF THE AMERICAN PHARMACEUTICAL ASSOCIATION
HONORARY MEMBER OF THE PHILADELPHIA COLLEGE OF PHARMACY
PROFESSOR OF BOTANY IN KING'S COLLEGE, LONDON
PROFESSOR OF BOTANY AND MATERIA MEDICA TO THE PHARMACEUTICAL SOCIETY OF GREAT BRITAIN

LONDON
J. & A. CHURCHILL
11 NEW BURLINGTON STREET
1884

TO

HENRY TRIMEN, M.B., F.L.S.

Director of the Royal Botanic Gardens, Pérádeniya, Ceylon

A FORMER PUPIL

WHO EARLY DISTINGUISHED HIMSELF AS A BOTANIST

AND WHO FOR MORE THAN SIX YEARS WAS ASSOCIATED WITH

THE AUTHOR IN BRINGING OUT THEIR WORK ON

'MEDICINAL PLANTS'

WHICH OWES SO MUCH OF ITS SUCCESS TO HIS

ENERGY, PERSEVERANCE, AND ABILITY

This little work is Dedicated

IN REMEMBRANCE OF MANY YEARS OF

CO-OPERATION AND FRIENDSHIP

AND WITH EVERY FEELING OF REGARD AND ESTEEM

BY HIS SINCERE FRIEND

THE AUTHOR

PREFACE.

The present little work is issued in accordance with the promise made by the author last summer, when his 'Student's Guide to Structural, Morphological, and Physiological Botany' was published, and to which volume it forms a necessary supplement. Its publication has also afforded the author the long desired opportunity of carrying out the intention which he expressed as far back as 1861, when the first edition of his 'Manual of Botany' appeared, namely, of including a short notice of Descriptive Botany. It is believed that the present work cannot but form a most convenient and handy little volume for use abroad and at home by medical, pharmaceutical, and other students who are desirous of obtaining a good practical knowledge of some of the more important British natural orders and their medicinal plants, and also as a foundation for further study.

All the orders required by the various examining bodies in this country have been included; and as their characters have been very fully illustrated, no difficulty ought to be experienced by the careful student in becoming practically acquainted with them by the aid of such illustrations and the examination of, at the same time, fresh specimens of plants. The study of plants in the fields and gardens, and by careful dissection of specimens at home, is, indeed, as necessary to those who would be practically acquainted with Botany, as work in a chemical laboratory is to those who would desire to acquire a practical knowledge of Chemistry.

April 1884.

CONTENTS.

BOOK II.

DESCRIPTIVE BOTANY.

CHAPTER I.

Section 1.

THE STUDENT'S GUIDE

TO

SYSTEMATIC BOTANY.

WE have in our last little work, entitled 'The Student's Guide to Morphological, Structural, and Physiological Botany,' treated of the forms, structure, and special functions of the various organs of plants, and also of the life of the whole plant. In thus directing our attention to these departments of Botany, we cannot but have noticed the almost infinite varieties of forms which have been presented to us, and also at the same time observed that, notwithstanding such variations, there are some striking resemblances in the structure of the organs of certain plants, by which a close relationship is thus clearly indicated between them. It is the object of Systematic Botany to take notice of such relationships, and thus to bring plants together which are allied in their forms and structure, and to separate those that are unlike; and in this way to take a comprehensive view of the whole Vegetable Kingdom. In its extended sense, Systematic Botany has for its object the naming, describing, and arranging of plants in such a manner, that we may readily ascertain their names, and at the same time get an insight into their affinities and general properties.

In this little work we can only treat very generally and briefly of the subjects to be described, and it will be most convenient and intelligible to do so under the two heads of Classification of Plants and Descriptive Botany.

BOOK I.

CLASSIFICATION OF PLANTS.

CHAPTER I.

GENERAL PRINCIPLES OF CLASSIFICATION.

Section 1. Species, Genera, Orders, Classes, and Sub-kingdoms.

At the present time there are at least 120,000 species of plants known to exist on the earth. It is absolutely necessary, therefore, for the purpose of study, or in order to obtain any satisfactory knowledge of such a vast number of individuals, that we should arrange them according to some definite and fixed rules; but before we proceed to describe the systems that have been devised at various times for their arrangement, it will be necessary to define certain terms which are in common use in such systems.

1. Species.—By the term species we understand a collection of individuals which resemble each other more nearly than they resemble any other plants, and which, taking Flowering Plants as illustrations, can be reproduced by seed; so that we may from analogy infer that they have all been derived from a common stock. Thus, if we walk into a field of Beans, Peas, or Clover, we observe thousands of individuals which, although differing to a certain extent in size, and in some other unimportant characters, we at once associate together under a common name. In like manner we commonly observe around us, in the gardens and fields,

similar collections of individuals. Such collections of plants, thus seen to resemble each other in all their important parts, constitute our first idea of a species; and that idea is at once confirmed if, by taking the seeds of such plants and sowing them, we obtain other plants resembling those from which such seeds have been derived. Species are, however, under special conditions, liable to variations, and we have then formed what are termed *varieties* and *races*.

a. *Varieties.*—It has just been observed that if the seed of a species be sown, it will reproduce its parent, or, in other words, produce a plant resembling its parent in all its important parts. But this will only happen when the new individual has been exposed to similar influences of soil, heat, light, moisture, and other conditions, as its parent; and hence we find that variations in such particulars will lead to certain peculiarities in form, colour, size, and other minor characters, in plants raised from the seeds of the same species. In this manner we have produced what are termed *varieties*. In some cases such variations are merely transient, and the individuals presenting such peculiarities will in time return to their original specific type, or perish altogether; while in other instances they are permanent and continue throughout the life of the individual, the whole plant being, as it were, impregnated with the particular variations thus impressed upon it, and hence such variations may be perpetuated by the gardener as in the operations of Budding and Grafting, as is the case with many of our fruit trees and flowers. But even these varieties cannot be propagated by seed; for if their seeds be sown, the individuals which will be produced will have a tendency to revert to the original species from which such varieties have been obtained, so that the nature of the plant raised will depend upon the character of the soil in which it is placed, and the other external conditions to which it is exposed. Thus, if we sow the seeds of a number of different varieties of Apples, the fruit which will be afterwards produced by the new generation of Apple trees will, instead of resembling that of their parents, have a tendency to revert to that of the common Crab, from which species all such varieties have been originally derived. Hence a variety differs essentially from a species in the fact that it cannot be propagated by seed.

b. *Races.*—Besides the varieties just alluded to there are others, which are called *permanent varieties* or *races*, because their peculiarities can be transmitted by seed. Familiar examples of such races are afforded by our Cereal grains, as Wheat, Oats, and Barley; and also by our culinary vegetables, as Peas, Lettuce, Radishes, Cabbages, Cauliflowers, and Broccoli. How such races of plants have originated, it is impossible to say with any certainty. In the first case they probably arose in an accidental manner, for it is found that plants under cultivation are liable to produce certain variations or abnormal deviations from their specific type, or to *sport*, as it is termed. By further cultivation under the care of the gardener, such variations are after a time rendered permanent, and can be propagated by seed. These so-called permanent varieties, however, if left to themselves, or if sown in poor soil, will soon lose their peculiarities, and either perish or return to their original specific type; it will be seen, therefore, that races present well-marked characters by which they are distinguished from true species. Hence, although our cereal grains and culinary vegetables have become permanent varieties by ages of cultivation and by the skill of the cultivator, they can only be made to continue in that state by a resort to the same means, for if left to themselves they would, as just observed, either perish or revert to their original specific type; and hence we see also how important is the assistance of the agriculturist and gardener in perpetuating and improving such variations.

Another cause which leads to constant variations from the specific type, is *hybridisation*. The varieties thus formed, which are called *hybrids* or *cross-breeds*, are, however, rarely transmitted by seed—although, in some instances, such is the case for a few generations—but they gradually revert to one or the other parent stock.

2. GENERA.—The most superficial observer of plants will have noticed that certain species are more nearly allied to each other than to other species. Thus, the different kinds of Roses, Brambles, Heaths, and Willows, may be cited as familiar examples of such assemblages of species; for, although the plants comprehended under these names present certain well-marked distinctive characters, yet there are at the same time also striking resemblances between them. Such assemblages of species are called *genera*. A *genus*, therefore, is a collection of species which resemble each other in general structure and appearance more than they resemble any other species. Thus, the various kinds of Brambles constitute one genus, the Roses another, the Willows, Heaths, Clovers, and Oaks form also, in like manner, as many different genera. The characters of a genus are taken *exclusively* from the organs of reproduction, while those of a species are derived *generally* from all parts of the plant; hence a genus is defined as a collection of species which resemble each other in the structure and general characters of their organs of reproduction. It is not necessary, however, that a genus should contain a number of species, for, if a single species presents peculiarities of a marked kind, it may of itself constitute a genus.

It frequently happens that two or more species of a genus have a more striking resemblance to each other in certain important characters than to other species of the same genus; in which case they are grouped together into what is termed a *sub-genus*.

3. ORDERS.—If we regard collections of genera from the same point of view as we have just done those of species,—that is, as to their resemblance or family likeness,—we shall find that some of them also resemble each other

more than they do other genera. Thus, Turnips and Radishes have a strong common resemblance, while they are unlike Strawberries and Brambles; and even less so to Oaks and Beeches; and still more unlike Larches and Pines. Proceeding in this way throughout the Vegetable Kingdom, we collect together allied genera, and form them into groups of a higher order called *Orders* or *Natural Orders*; hence, while genera are collections of related species, orders are collections of allied genera. Thus Turnips and Radishes belong to different genera, but they agree in their general structure, and are hence included in one order, which is termed *Cruciferæ*; while Strawberries, Brambles, Cinquefoils, Roses, Apples, and Plums, are all different genera, but from the general resemblance they bear to each other in their structure, they are placed in one order, called *Rosaceæ*. Again, Oaks and Beeches are different genera, but they belong to one order; also the Larches and Pines are different genera, but as the fruit of them all is a *cone*, they are grouped together in one order, which is termed the *Coniferæ*.

We find also that certain genera of an order, like certain species of a genus, have a more striking resemblance to each other than to other genera of the same order; hence such are grouped together into what are called *Sub-orders*. Thus the Chicory, Dandelion, Lettuce, Thistle, Burdock, and Chamomile, all belong to the same order, but there is a greater resemblance in the Chicory, Dandelion, and Lettuce to each other than to the remaining genera. Hence, while all the above genera belong to the order *Compositæ*, they are at the same time placed in two different sub-orders. Thus, one sub-order, called the *Liguliforæ*, includes the Chicory, Dandelion, and Lettuce; and the other sub-order, termed the *Tubuliforæ*, that of the Thistle, Burdock, and Chamomile. In like manner, while we find the Almond, Plum, Strawberry, Rose, Medlar, and Apple, all belonging to the same order *Rosaceæ*, some of them have more resemblance to each other than to other plants in the same order. Thus, the Almond and Plum have a drupaceous fruit, and are therefore placed in a distinct sub-order, which is called *Amygdaleæ*; the Strawberry and Rose are much more like each other than they are to the Almond and Plum, or to the Apple and Medlar, hence they are put in a sub-order called *Roseæ*; while the Apple and Medlar, from the character of their fruit, are placed in a sub-order termed *Pomeæ*.

It is also found convenient at times to subdivide sub-orders into *Tribes* and *Sub-tribes*, by collecting together into groups certain very nearly allied genera; but it is not necessary for us to illustrate such divisions further, as the principles upon which they depend have been now sufficiently treated of.

4. CLASSES.—By a class we understand a group of orders possessing some important structural characters in common. Thus we have the classes Monocotyledones, Dicotyledones, and Acotyledones, which possess certain distinctive characters in their embryos, from which they derive their names; as well as other important anatomical differences.

The Classes are also divided into *Sub-classes* and other divisions, in the same manner as the orders and genera are thus subdivided; but as such divisions vary in different systems, and are all more or less artificial, it is not necessary for us, in this place, to dwell upon them further.

5. SUB-KINGDOMS.—In many natural systems, as in that adopted in this work, the classes are also arranged in higher divisions called Sub-kingdoms, which, as their name implies, are the primary divisions of the Vegetable Kingdom. These are characterised by some highly important structural and very general distinctive peculiarities, as the presence or absence of flowers, and whether forming seeds or spores.

Section 2. Nomenclature.

It is the object of nomenclature to lay down rules for naming the various kinds of plants and the different groups into which they are arranged in our systems of classification; in the same manner as it is the object of terminology to find names for the different organs of plants and the modifications which those organs present.

1. Species.—The names of the species are variously derived. Thus the species of the genus *Viola*, as shown by Gray in the following paragraphs, exhibit the origin of many such names. 'Specific names sometimes distinguish the country which a plant inhabits, for example, *Viola canadensis*, the Canadian Violet; or the station where it naturally grows, as *Viola palustris*, which is found in swamps, and *Viola arvensis*, in fields; or they express some obvious character of the species, as *Viola rostrata*, where the corolla bears a remarkably long spur, *Viola tricolor*, which has tri-coloured flowers, *Viola rotundifolia*, with rounded leaves, *Viola lanceolata*, with lanceolate leaves, *Viola pedata*, with pedately-parted leaves, *Viola primulæfolia*, where the leaves are compared to those of a Primrose, *Viola asarifolia*, where they are likened to those of *Asarum*, *Viola pubescens*, which is hairy throughout, &c. Frequently the species bears the name of its discoverer or describer, as *Viola Muhlenbergii*, *Viola Nuttallii*, &c.'

Specific names are written after the generic, as indicated above in the different species of the genus *Viola*, and these together constitute the proper appellation of a plant, in the same way as the surnames and Christian names designate the members of a family. The specific names should also in all cases be adjectives, or substantives used adjectively; in the former case they should agree in gender and case with the name of the genus. Thus when a species is named after its discoverer or describer, it is usually placed in the genitive case, as *Viola Muhlenbergii* and *V. Nuttallii*; but when such names are merely given in honour of botanists who have had nothing to do with their discovery or description, the specific names are generally put in the adjective form, as *Carex Hookeriana*, *Veronica Lindleyana*: such a rule is, however, frequently departed from. Sometimes the specific name is a noun, in which case it does not necessarily agree with the genus in gender; such specific names are often old generic ones, as *Dictamnus Fraxinella*, *Rhus Cotinus*, *Lythrum Salicaria*, *Rhus Coriaria*, *Dianthus Armeria*, *Asclepias Vincetoxicum*. In such cases the specific name should begin with a capital letter: a similar rule should also be adopted when it is derived from a person; but in all other instances it is better that the specific name should begin with a small letter. The specific name was called by Linnæus the *trivial* name; thus, in the particular kind of Violet called *Viola palustris*, *Viola* is the generic, and *palustris* the specific or trivial name.

2. GENERA.—The names of the genera are substantives in accordance with the rule laid down by Linnæus as follows:—Every species shall have a particular name, compounded of a substantive and an adjective, whereof the former indicates the genus, and the latter the species. This has already been referred to under the head of Species. The names of the genera are derived in various ways: thus, either from the name of some eminent botanist, as Linnæa after Linnæus, Smithia after Smith, Hookeria after Hooker, Jussiæa after Jussieu, Tournefortia after Tournefort; or from some peculiarity of structure or habit of the plants comprised in them, and from various other circumstances. Thus, *Crassula* is derived from the genus comprising plants with succulent or thickened leaves; *Sagittaria*, from its arrow-shaped leaves; *Arenaria*, from growing in sandy places; *Lithospermum*, from its fruits, which were formerly regarded as seeds, having a

stony hardness; *Campanula*, from its corolla being in the form of a bell; *Lactuca*, from its milky juice; and so on. Others, again, have derived their generic names from supposed medicinal properties, such as *Scrophularia*, from its former use in scrofula; *Pulmonaria*, from its employment in pulmonary disease, &c.

3. ORDERS. — The names of the Orders in the Artificial System of Linnæus are chiefly derived from the various characters of the gynœcium and fruit. Those of Natural Systems are usually taken from some well-known genus which is included in any particular order, and which may be regarded as the type of that order. Thus, the genus *Ranunculus* gives the name *Ranunculaceæ* to the order to which it belongs; the genera *Papaver*, *Malva*, *Hypericum*, *Geranium*, *Rosa*, *Lilium*, *Orchis*, and *Iris*, in like manner give names respectively to the orders *Papaveraceæ*, *Malvaceæ*, *Hypericaceæ*, *Geraniaceæ*, *Rosaceæ*, *Liliaceæ*, *Orchidaceæ*, and *Iridaceæ*. At other times the names of the orders are derived from some characteristic feature which the plants included in them present. Thus, the order *Cruciferæ* is so named, because its plants have cruciate corollas; the order *Leguminosæ* comprises plants whose fruit is a legume; the *Umbelliferæ* are umbel-bearing plants; the *Labiatæ* have a labiate corolla; the *Coniferæ* are cone-bearing plants; and so on.

4. CLASSES.—The names of the classes are derived from some important and permanent characters which the plants comprised in them possess, relating either to their structure or mode of development. Such names vary, however, according to the views of different systematic botanists. Those more commonly used in this country, and which have been accordingly adopted in this work, are, *Acotyledones*, *Monocotyledones*, and *Dicotyledones*,—terms which are derived from the structure of the embryo in the three classes respectively. Others, also in common use, are derived from the absence or presence of a stem, and its mode of develop-

ment and structure: such are *Exogens*, *Endogens*, *Acrogens*, and *Thallogens*. The above names are used especially in Natural Systems of Classification; while the names of Classes in the Artificial System of Linnæus are derived chiefly from the number and other characters presented by the andrœcium.

5. SUB-KINGDOMS.—The names of these are generally derived from the presence or otherwise of evident flowers or reproductive organs, as those of *Phanerogamia* and *Cryptogamia*. The names of *Cotyledones* and *Acotyledones*, indicating the presence or absence of an embryo, are also in common use. Others, again, have been employed, having reference to their elementary structure, as *Vasculares* and *Cellulares*; or to the presence or absence of a stem, as *Cormophyta* and *Thallophyta*.

CHAPTER II.

SYSTEMS OF CLASSIFICATION.

WE have already stated that Systematic Botany has for its object the naming, describing, and arranging of plants in such a manner that we may readily ascertain their names, and at the same time get an insight into their affinities and general properties. Every system that has been devised for the arrangement of plants does not, however, comprise all the above points; for, while some systems are of value simply for affording us a ready means of ascertaining their names, others not only do this, but at the same time give us a knowledge of their affinities and properties. Hence we arrange the different systems of Classification under two heads; namely, Artificial and Natural,—the former only necessarily enabling us to ascertain readily the name of a particular plant; while the latter, if perfect, should comprise all the points which come within the object of Systematic Botany. The great aim of the botanist, therefore, should be the development of a true Natural System; but, in past times, artificial systems, more particularly that of Linnæus, have been of great value. Linnæus himself never devised his system with any expectation or desire of its serving more than a temporary purpose, or as an introduction to the Natural System, when the materials for its formation had been obtained.

In both artificial and natural systems, the lower divisions—namely, the species and genera, are the same, the difference between the systems consisting in the manner in which these divisions are grouped into orders, classes, and other higher groups. Thus, in the Linnæan and other Artificial Systems, one, or, at most, a few characters, are arbitrarily selected, and all the plants in the Vegetable Kingdom are distributed under classes and orders according to the correspondence or difference of the several genera in such respects, no regard being had to any other characters. The plants in the classes and orders of an artificial system have, therefore, no necessary agreement with each other except in the characters selected for convenience as the types of those divisions respectively. Hence such a system may be compared to a dictionary, in which words are arranged, for convenience of reference, in an alphabetical order, adjacent words having no necessary agreement with each other, except in commencing with the same letter. In the Natural System, on the contrary, all the characters of the genera are taken into consideration, and those are grouped together into orders which correspond in the greatest number of important characters; and the orders are again united, upon the same principles, into groups of a higher order, namely, the classes and sub-kingdoms. While it must be evident, therefore, that all the knowledge we necessarily gain by an artificial system, is the name of an unknown plant; on the other hand, by the natural system, we learn not only the name, but also its relations to the plants by which it is surrounded, and hence get a clue to its structure, properties, and history. Thus, supposing we find a plant, and wish to ascertain its name, if we turn to the Linnæan system, and find that such a plant is the *Menyanthes trifoliata*, this name is the whole amount of the knowledge we have gained; but by turning to the Natural System instead, and finding that our plant belongs to the order *Gentianaceæ*, we ascertain at once from its affinities that it must have the tonic and other properties which are possessed by the plants generally of that order, and, at the same time, we also learn that it accords in its structure with the same plants; and hence, by knowing the name of a plant by the Natural System, we may at once learn all that is most important in its history. It is quite true that all the orders, as at present constituted, are by no means so natural as that of the *Gentianaceæ*, but this arises from the present imperfection of our systems, and can only be remedied as our knowledge of plants extends; even a system, devised as perfectly as possible one day, may be deficient the next, in consequence of new plants being discovered which might compel us to alter our views, for at present the Floras of some regions of the globe are imperfectly, and of others, almost entirely unknown. Sufficient, however, is known of plants to enable us to establish certain great divisions according to a natural method, and which after discoveries are not likely to affect to any important extent. The present imperfections of the Natural System are, therefore, comparatively unimportant, and will no doubt disappear as our knowledge of the Flora of the globe becomes extended.

In my 'Manual of Botany' the more important artificial and natural systems that have been or are now in use in this and other countries have been fully noticed (see pages 401–419). But in the present work it would be out of place to do more than describe generally the Natural System we have here adopted, and which, in its essential particulars, is that commonly in use in this country. This is founded upon the systems of Jussieu, De Candolle, and Lindley,—that of De Candolle being the basis. It is as follows:—

The Vegetable Kingdom is first divided into two sub-kingdoms, namely:—Phanerogamia, Flowering, or Cotyledones; and Cryptogamia, Flowerless, or Acotyledones.

Sub-kingdom 1. *Phanerogamia.*—This includes plants which have evident flowers; and which are reproduced by seeds containing an embryo with one or more cotyledons.

Sub-kingdom 2. *Cryptogamia.*—This contains those plants which have no flowers; and which are reproduced by spores, and are therefore acotyledonous.

The *Phanerogamia* is divided into two classes, and other sub-divisions, thus:—

Class 1. DICOTYLEDONES, in which the embryo is dicotyledonous; the germination exorhizal; the stem exogenous; the leaves with a reticulated venation; and the flowers with a quinary or quaternary arrangement. In this class we have two divisions.

Division 1. *Angiospermia*, in which the ovules are enclosed in an ovary; and which are fertilised indirectly by the action of the pollen on the stigma. In this division we have four sub-classes:—

Sub-class 1. *Thalamifloræ.*—This comprises plants with flowers which are usually furnished with both a calyx and corolla; the latter composed of distinct petals inserted on the thalamus; stamens hypogynous, or adherent to the sides of the ovary, that is, arising directly from the thalamus, or placed on the outside of an hypogynous disk.

Sub-class 2. *Calyciflorœ.*—Flowers having usually a calyx and corolla; the latter mostly with distinct petals, and inserted on the calyx or a perigynous disk; stamens either perigynous or epigynous. This sub-class has two sub-divisions:—

1. *Perigynœ*, in which the calyx is free, or nearly so; the stamens usually perigynous; and the ovary superior.
2. *Epigynœ*, in which the calyx is more or less adherent; the stamens epigynous; and the ovary inferior.

Sub class 3. *Corolliflorœ.*—Flowers having both a calyx and corolla; the latter with united petals; stamens inserted on the corolla or ovary, or free and arising from the thalamus. Of this sub-class we have three sub-divisions:—

1. *Epigynœ*, in which the calyx is adherent; and the ovary consequently inferior.
2. *Hypostamineœ*, in which the stamens are inserted on the thalamus, and do not adhere to the corolla; and the ovary superior.
3. *Epipetalœ*, in which the corolla arises from the thalamus, and has the stamens attached to it; and the ovary superior.

Sub-class 4. *Monochlamydeœ* or *Incompletœ.*—Flowers either having a calyx only, or without both calyx and corolla.

Division 2. *Gymnospermia*, in which the ovules are naked or not enclosed in an ovary, and which are fertilised directly by the action of the pollen.

Class 2. MONOCOTYLEDONES, in which the embryo is monocotyledonous; the germination endorhizal; the stem endogenous; the leaves usually with a parallel venation; and the flowers with a ternary arrangement. In this class we have two sub-classes.

Sub-class 1. *Petaloideæ.*—Leaves with a parallel venation, or rarely reticulated; floral envelopes (perianth) verticillate and usually coloured, or rarely scaly, and sometimes absent. This sub-class has three sub-divisions:—

1. *Epigynæ*, in which the flowers are usually hermaphrodite; the perianth adherent; and the ovary inferior.
2. *Hypogynæ*, in which the flowers are usually hermaphrodite; the perianth free; and the ovary superior.
3. *Diclines*, in which the flowers are usually unisexual; and the perianth either absent or consisting of a few scales.

Sub-class 2. *Glumaceæ* or *Gluміferæ.* — Leaves parallel-veined; flowers glumaceous, that is, having no proper perianth, but imbricated bracts instead.

The *Cryptogamia* constitutes a class by itself, thus:—

Class 3. ACOTYLEDONES, in which the plants are reproduced by spores, and are therefore acotyledonous, and have an heterorhizal germination; the stem is present or absent, in the former case, when woody, it is acrogenous; the leaves or fronds are also either absent or present, in which latter case the veins are commonly forked;

they have no true flowers. This has two sub-classes:—

Sub-class 1. *Acrogenæ* or *Cormophyta.*—Plants with the stem and leaves or fronds distinguishable.

Sub-class 2. *Thallogenæ* or *Thallophyta.*—Plants with no distinction of stem and leaves.

The following is a tabular arrangement of the above system:—

VEGETABLE KINGDOM.

Sub-kingdom 1. Phanerogamia, Cotyledones, or Flowering Plants.

Class 1. DICOTYLEDONES.
Division 1. Angiospermia.
Sub-class 1. Thalamifloræ.
2. Calycifloræ.
1. Perigynæ.
2. Epigynæ.
3. Corollifloræ.
1. Epigynæ.
2. Hypostamineæ.
3. Epipetalæ.
4. Monochlamydeæ or Incompletæ.
Division 2. Gymnospermia.

Class 2. MONOCOTYLEDONES.
Sub-class 1. Petaloideæ.
1. Epigynæ.
2. Hypogynæ.
3. Diclines.
2. Glumaceæ or Glumiferæ.

Sub-kingdom 2. Cryptogamia, Acotyledones, or Flowerless Plants.

Class 3. ACOTYLEDONES.
Sub-class 1. Acrogenæ or Cormophyta.
2. Thallogenæ or Thallophyta.

CHAPTER III.

ARRANGEMENT AND CHARACTERS OF THE NATURAL ORDERS.

HAVING now given a description of the natural system adopted in this volume, we proceed to illustrate it by describing a few of the more important natural orders, containing British plants, which have been arranged under its various sub-divisions.

SUB-KINGDOM I.

PHANEROGAMIA, COTYLEDONES, OR FLOWERING PLANTS.

CLASS I. DICOTYLEDONES.

Division I. ANGIOSPERMIA.

Sub-class I. *Thalamifloræ.*

RANUNCULACEÆ, the Crowfoot Order.—Character.—*Herbs*, or rarely climbing *shrubs*, with a watery, colourless, and usually acrid, juice. *Leaves* alternate or opposite, generally much divided (*fig.* 1), or sometimes entire, with usually enlarged and amplexicaul petioles. *Stipules* generally absent, but rarely present, and then adnate. *Inflorescence* definite (*fig.* 1), or indefinite. *Calyx* of 3—6, usually 5 (*fig.* 2), distinct sepals, regular (*figs.* 1 and 2), or irregular (*fig.* 4), green or rarely petaloid, deciduous or very rarely persistent (*fig.* 11); æstivation generally imbricate (*fig.* 2), sometimes valvate (*fig.* 3) or induplicate. *Corolla* of 3—15, usually 5 (*figs.* 1 and 2), distinct petals, regular (*figs.* 1 and 2), or irregular (*figs.* 4 and 5) æstivation imbricate (*figs.* 1 and 2), sometimes absent (*fig.* 3). *Stamens* numerous (*figs.* 2, 3, and 5), or very rarely few, hypogynous (*fig.* 6, *e*); *anthers* adnate (*fig.* 7), bursting longitudinally. *Carpels* generally numerous (*figs*. 2, 3, and 6, *p*), rarely few or 1, usually distinct and one-celled (*fig*. 10, *o*), or very rarely united below so as to form a compound many-celled ovary; *ovary* with one (*fig*. 10, *g*) or many, anatropous ovules. *Fruit* either consisting of a number of dry achenes, or of one or more whorls of follicles (*fig*. 8), which are sometimes united below, or sometimes the follicle is

FIG. 1.

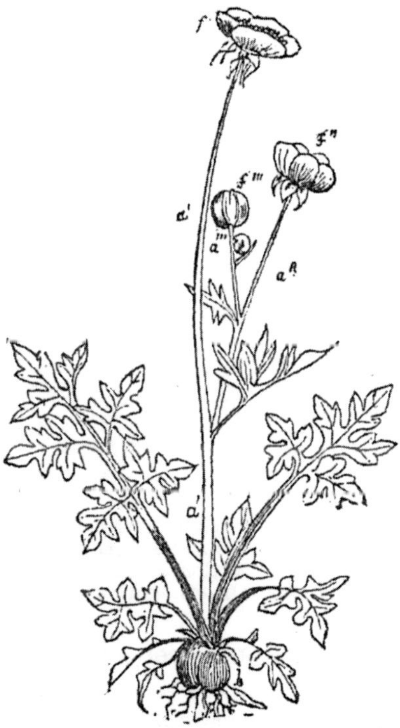

FIG. 2.

FIG. 3.

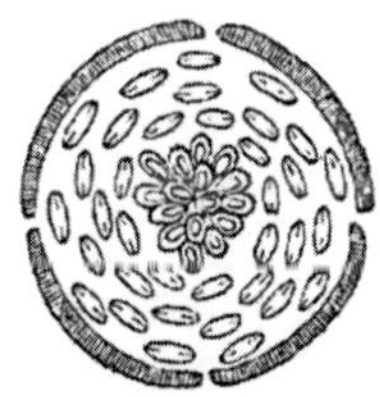

Fig. 1. A plant of *Ranunculus bulbosus*.——*Fig*. 2. Diagram of the flower of a species of Ranunculus.——*Fig*. 3. Diagram of the flower of a species of Clematis.

solitary; or very rarely the fruit is baccate, with one or more seeds. *Seeds* solitary or numerous, when solitary, erect or pendulous; *embryo* minute (*fig*. 9, *emb*), at the base of homogeneous horny albumen, *alb*.

Diagnosis.—Herbs or rarely shrubs, with a colourless, watery, and usually acrid, juice. Sepals, petals, and stamens distinct, hypogynous. Corolla with an imbricate æstivation. Stamens usually numerous; anthers adnate, bursting longitudinally. Carpels, except in a very few instances, more or less distinct. Seeds with a minute embryo, and homogeneous horny albumen, anatropous.

Division of the Order and Illustrative Genera.—The order may be divided into five tribes as follows:—

Tribe 1. *Clematideæ.* Calyx valvate (*fig.* 3), or indu-

FIG. 4.

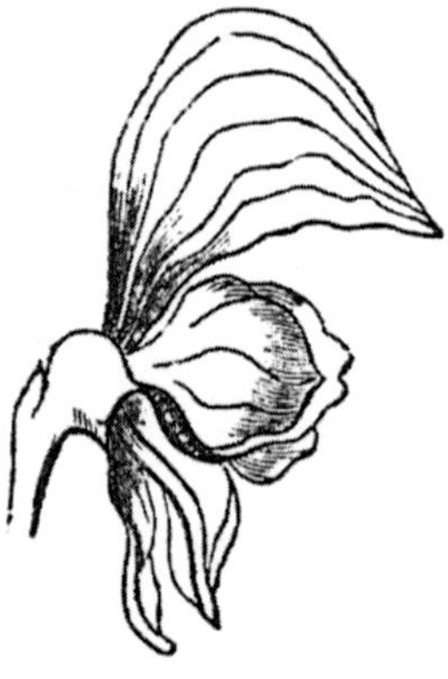

FIG. 5.

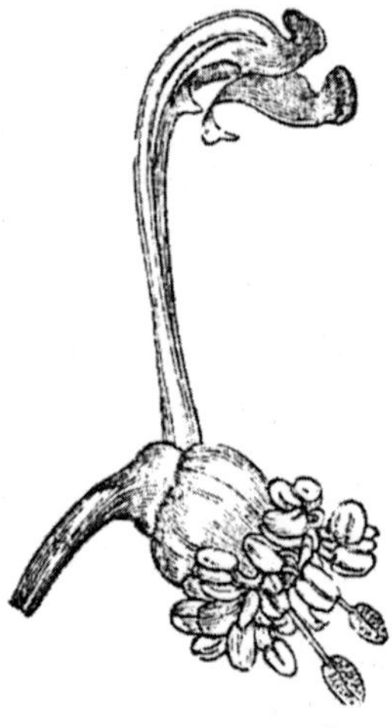

Fig. 4. Flower of Monkshood (*Aconitum Napellus*), with an irregular polysepalous calyx. The upper sepal is petaloid, and hooded or helmet-shaped.——*Fig.* 5. A portion of the flower of the Monkshood, with numerous stamens below, and two stalked somewhat horn-shaped petals above.

plicate. Fruit consisting of a number of achenes. Seed pendulous. *Illustrative Genus*:—Clematis, *Linn.*

Tribe 2. *Anemoneæ.* Calyx imbricate, usually coloured. Fruit consisting of a number of achenes. Seed pendulous. *Illustrative Genus*: Anemone, *Haller.*

Tribe 3. *Ranunculeæ.* Calyx imbricate (*fig.* 2). Fruit consisting of a number of achenes. Seed erect. *Illustrative Genus*:—Ranunculus, *Linn.*

Tribe 4. *Helleboreæ.* Calyx imbricate. Petals irregular (*figs.* 4 and 5) or none. Fruit consisting of one or more whorls of many-seeded follicles (*fig.* 8), which are sometimes united below; or rarely baccate. *Illustrative Genera* :—Aconitum, *Linn.*; Delphinium, *Linn.*; Actæa, *Linn.*

FIG. 6.

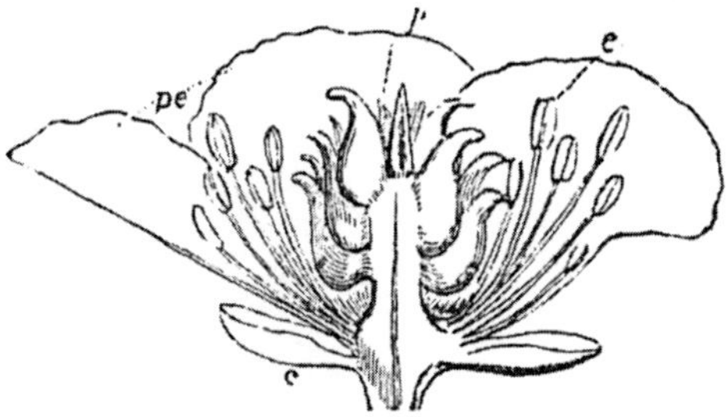

FIG. 7.

FIG. 8.

FIG. 9.

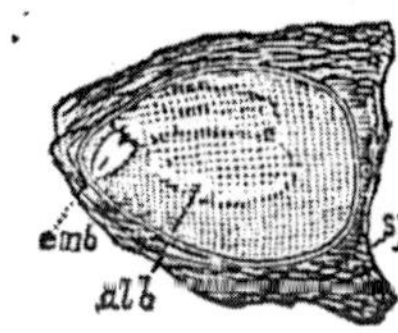

Fig. 6. Vertical section of the flower of *Ranunculus acris.* *c.* Calyx. *pe.* Petals. *e.* Stamens. *p.* Carpels.——*Fig.* 7. Adnate anther of a Ranunculaceous plant.——*Fig.* 8. Numerous follicles of *Trollius europœus.*——*Fig.* 9. Vertical section of the seed of the Monkshood (*Aconitum Napellus*). *sp.* Integuments of the seed. *emb.* Embryo. *alb.* Albumen.

FIG. 10.

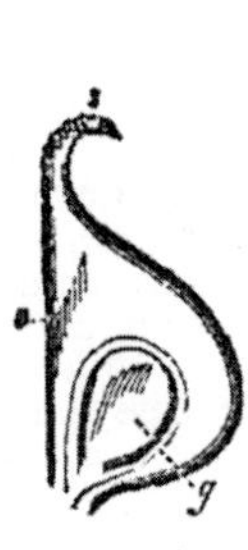

FIG. 11.

Fig. 10. Vertical section of a carpel of *Ranunculus acris.* *o.* Ovary. *g.* Ovule. *s.* Stigma.——*Fig.* 11. Pistil of the Tree Pæony (*Pæonia Moutan*) invested by a large cup-shaped expansion or disk, and having at its base a persistent calyx.

Tribe 5. *Pæonieæ.* Calyx imbricate. Fruit consisting of from 2–5 follicles, which are more or less surrounded by a cup-shaped hypogynous disk (*fig.* 11). *Illustrative Genus:*—Pæonia, *Linn.* There are no examples in British plants.

NYMPHÆACEÆ, the Water-Lily Order.—Character.

FIG. 12.

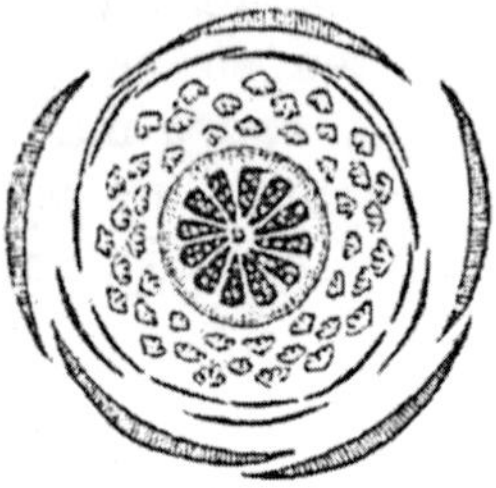

FIG. 13.

FIG. 14.

FIG. 15.

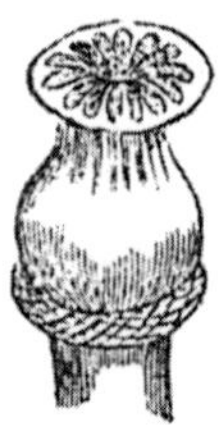

FIG. 16.

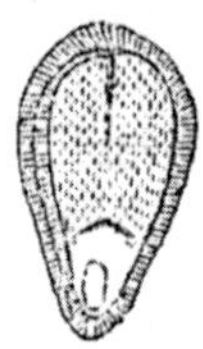

Fig. 12. Diagram of the flower of *Nuphar luteum.*——*Fig.* 13. Flower of the same.——*Fig.* 14. A portion of the flower of the White Water-lily (*Nymphæa alba*), consisting of a gynœcium invested by a large fleshy disk-like expansion prolonged from the thalamus below it. The ovary is surrounded by some stamens which have petaloid filaments and adnate introrse anthers, and by one petal and one sepal.——*Fig.* 15. Ovary of *Nuphar luteum* with numerous radiating stigmas.——*Fig.* 16. Vertical section of the seed of *Nymphæa alba*, showing the embryo enclosed in a vitellus, and on the outside of albumen.

—*Aquatic perennial herbs. Leaves* usually floating, peltate or cordate. *Flowers* solitary, large and showy. *Sepals* inferior, usually 4, or rarely 5 (*figs.* 12 and 13) or 6, persistent, generally petaloid on their inside. *Petals* numerous, deciduous, often passing by gradual transition into the stamens, in the same way as the sepals pass into the petals; inserted on a fleshy disk-like expansion of the thalamus below the stamens (*fig.* 14). *Stamens* numerous (*fig.* 12); *filaments* petaloid (*fig.* 14); *anthers* adnate. *Thalamus* large, forming a disk-like expansion more or less surrounding the ovary, and having inserted upon it the petals and stamens (*fig.* 14). *Carpels* numerous, united so as to form a compound ovary (*figs.* 14 and 15); *ovary* many-celled (*fig.* 12); *ovules* numerous, anatropous; *styles* absent; *stigmas* radiating on the top (*figs.* 13—15), and alternate with the dissepiments. *Fruit* indehiscent, many-celled. *Seeds* numerous, attached all over the spongy dissepiments; *embryo* minute, enclosed in a separate sac or vitellus, and on the outside of farinaceous albumen (*fig.* 16).

Diagnosis.—Aquatic perennial herbs with cordate or peltate usually floating leaves. Thalamus large, and forming a disk-like expansion more or less surrounding the ovary. Sepals inferior, persistent. Stamens numerous, with petaloid filaments and adnate anthers. Carpels united so as to form a compound many-celled ovary; stigmas radiating on the top, and alternate with the dissepiments; ovules numerous, and attached all over the dissepiments. Embryo minute, on the outside of farinaceous albumen, enclosed in a vitellus. *Illustrative Genera:*—Nymphæa, *Linn.*; Nuphar, *Smith.*

PAPAVERACEÆ, the Poppy Order.—Character.—*Herbs* or *shrubs*, usually with a milky juice (white or coloured). *Leaves* alternate, exstipulate. *Sepals* usually 2 (*figs.* 17 and 18), or rarely 3, caducous (*fig.* 17). *Petals* 4 (*figs.* 18 and 19), or rarely 6, or some multiple of 4, or very rarely wanting; usually crumpled in æstivation (*fig.* 18). *Stamens* generally numerous (*figs.* 18 and 19), hypogynous (*figs.* 19 and 20); *anthers* 2-celled, innate (*fig.* 20). *Ovary* 1-

celled, with 2 or more (*figs.* 18 and 21) parietal placentas, which project more or less from the walls into its cavity, and in *Romneya* actually adhere in the axis; *styles* absent (*fig.* 20) or very short; *stigmas* 2 (*fig.* 19, *sti*), or many (*fig.* 20, *sti*), alternate with the placentas, and opposite the imperfect dissepiments; when numerous, they form a star-like process on the top of the ovary (*fig.* 20); *ovules* numerous (*fig.* 21, *ov*). *Fruit* 1-celled, and either pod-shaped with 2 parietal placentas (*fig.* 22), or capsular with several

FIG. 17.

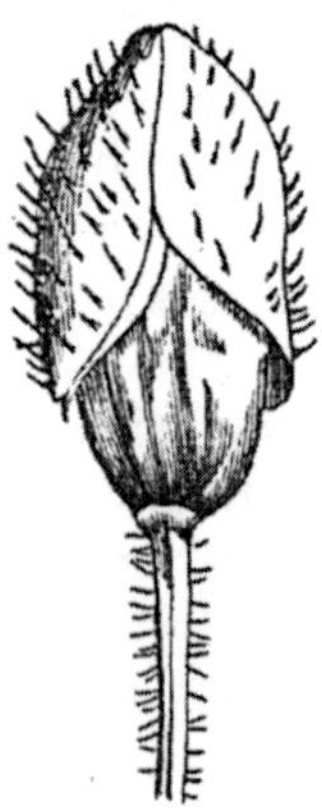

FIG. 18.

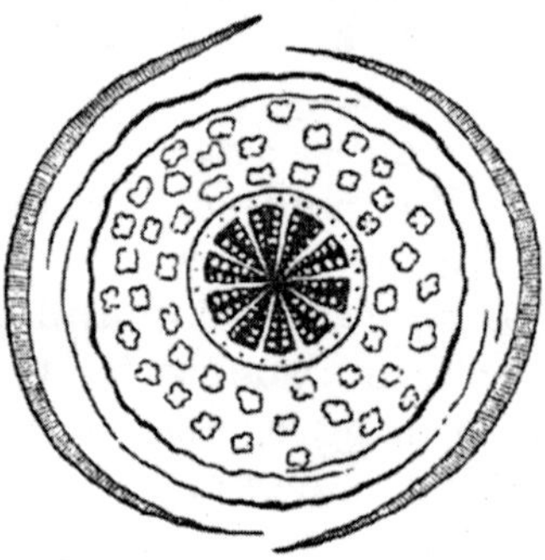

Fig. 17. Flower of Poppy, showing a caducous calyx.——*Fig.* 18. Diagram of the flower of the Poppy, with two sepals, four crumpled petals, numerous stamens, and a compound ovary with several parietal placentas.

placentas; dehiscing by valves (*fig.* 22), or pores, or sometimes indehiscent. *Seeds* usually numerous; *embryo* (*fig.* 23, *emb*) in fleshy-oily albumen, *alb*.

Diagnosis.—Usually herbs with a milky juice. Leaves alternate and exstipulate. Peduncles 1-flowered; flowers regular and symmetrical. Calyx and corolla with a binary or ternary arrangement of their parts, deciduous, hypogynous. Stamens usually numerous, hypogynous; anthers 2-celled, innate. Ovary compound, 1-celled, with 2 or many parietal placentas, stigmas alternate to the placentas. Fruit 1-celled.

Seeds numerous, albuminous. *Illustrative Genera*:—Papaver, *Linn.*; Chelidonium, *Linn.*

FIG. 19.

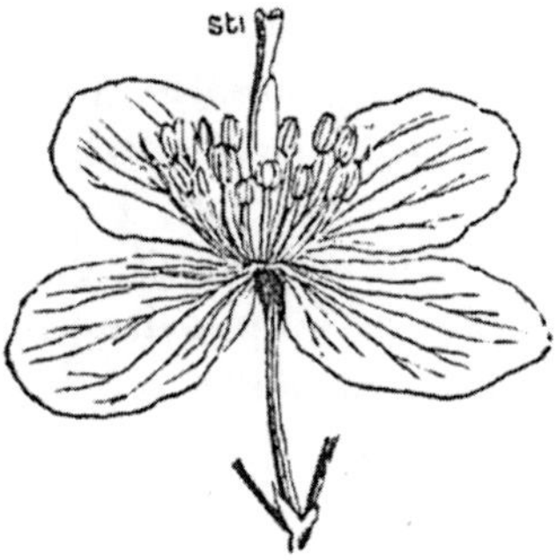

FIG. 20.

FIG. 22.

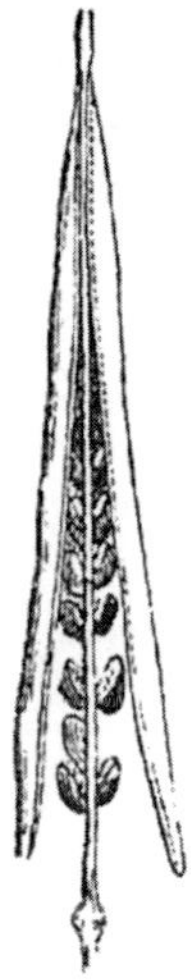

FIG. 21.

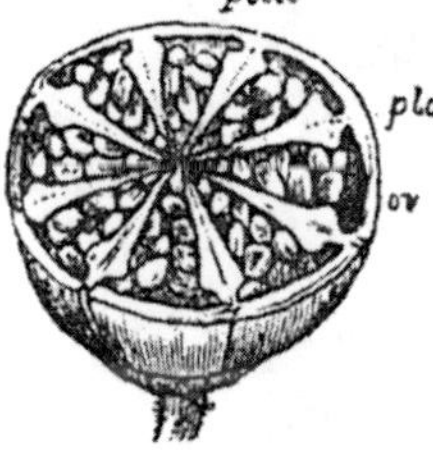

FIG. 23.

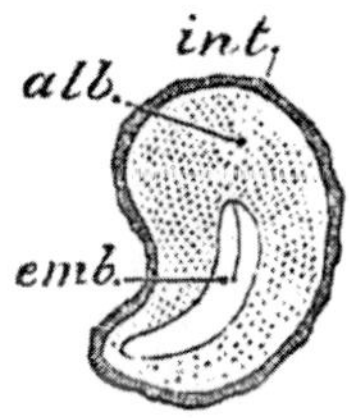

Fig. 19. Flower of Celandine (*Chelidonium majus*). *sti.* Two stigmas on the apex of a lengthened or pod-like ovary.——*Fig.* 20. Gynœcium of Poppy (*Papaver*), with one stamen arising from below it (*hypogynous*). *o.* United ovaries. *sti.* Stigmas.——*Fig.* 21. Transverse section of the ovary of a species of Poppy. *ov.* Ovules. *plac*, *plac.* Placentas, which in the young ovary nearly meet in the centre, and thus the ovary becomes almost multilocular, but as the ovary progresses in development it is distinctly unilocular.——*Fig.* 22. Siliquæform or pod-shaped capsule (*ceratium*) of Celandine.——*Fig.* 23. Vertical section of the seed of a species of Poppy (*Papaver*). *int.* Integuments. *emb.* Embryo. *alb.* Albumen.

CRUCIFERÆ, the Cruciferous Order.—Character.—*Herbs*, or very rarely shrubby plants. *Leaves* alternate, exstipulate. *Flowers* usually yellow or white, rarely purple, or some mixture of these colours; *inflorescence* indefinite, racemose (*fig.* 25) or corymbose; usually ebracteated. *Sepals* 4 (*fig.* 24), deciduous; *æstivation* imbricate (*fig.* 24), or rarely valvate. *Petals* 4 (*figs.* 24 and 25), hypogynous, arranged in the form of a Maltese cross, alternate with the sepals, deciduous. *Stamens* 6, tetradynamous (*fig.* 26, *ec*), hypogynous. *Thalamus* furnished with small green glands (*fig.* 26, *gl*) placed between the stamens. *Ovary* superior (*fig.* 26), with two parietal placentas (*fig.* 27), 1-celled, or more usually 2-celled (*fig.* 24) from the formation of a spurious dissepiment called the *replum* (*fig.* 27, *cl*); *style* none (*fig.* 26); *stigmas* 2 (*fig.* 26), opposite the placentas. *Fruit* a siliqua (*fig.* 28), or silicula (*figs.*

FIG. 24.

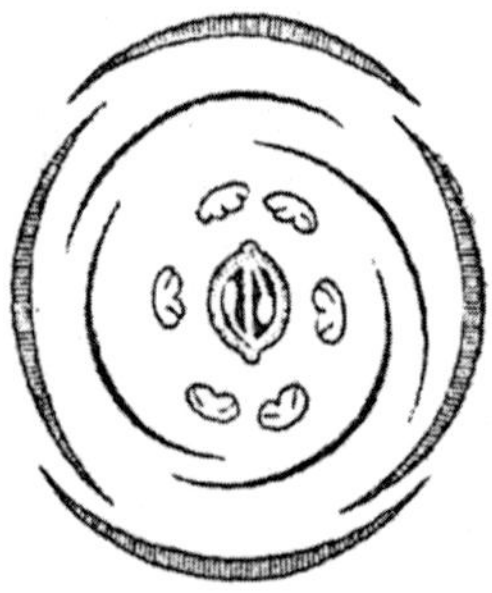

FIG. 25.

Fig. 24. Diagram of a Cruciferous flower.——*Fig.* 25. Portion of the flowering branch of the Wallflower (*Cheiranthus Cheiri*).

29 and 30), 1- or 2-celled, 1- or many-seeded. *Seeds* stalked, generally pendulous (*fig.* 29); *embryo* with the radicle variously folded upon the cotyledons (*figs.* 31–34); *albumen* none.

Diagnosis.—Generally ebracteated herbs. Inflorescence indefinite. Sepals and petals 4, deciduous, regular, the latter cruciate. Stamens hypogynous, tetradynamous. Ovary with 2 parietal placentas; stigmas 2. Fruit a siliqua or silicula. Seeds stalked, without albumen, and with the radicle variously folded upon the cotyledons. *No other order is likely to be confounded with this if ordinary care be taken, as tetra-*

dynamous stamens only occur here, except in a very few plants belonging to the natural order Capparidaceæ.

FIG. 26.

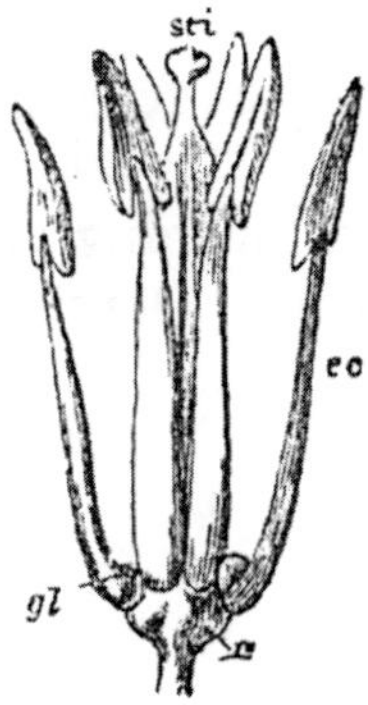

FIG. 27.

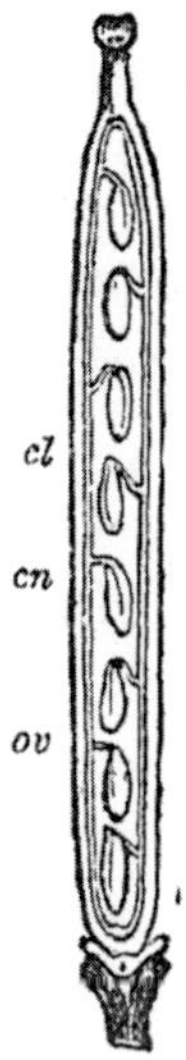

FIG. 28.

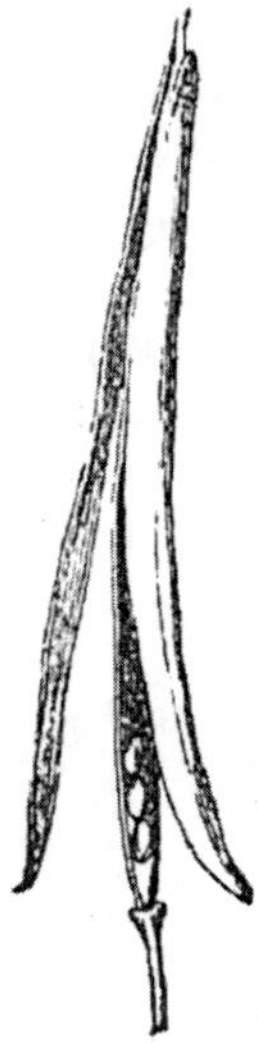

FIG. 29.

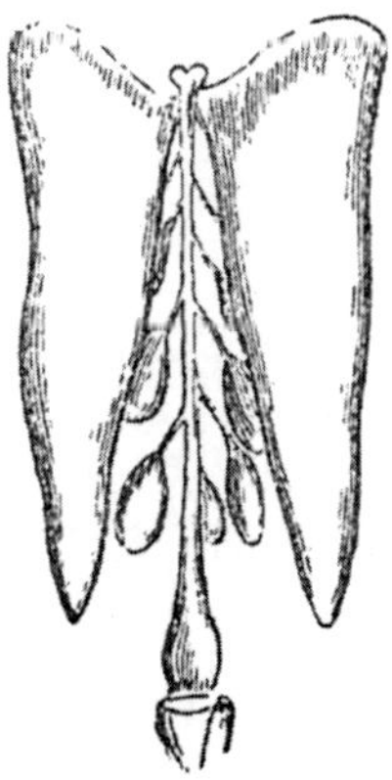

FIG. 30

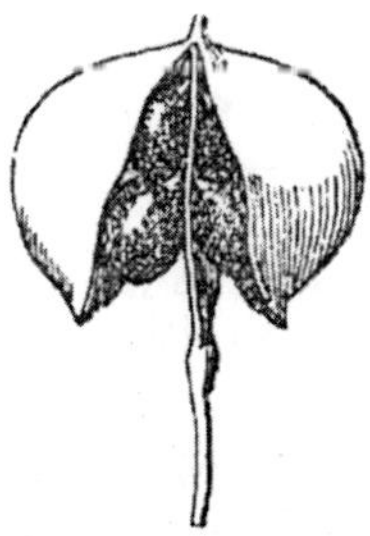

FIG. 31.

Fig. 26. Essential organs of the Wallflower (*Cheiranthus Cheiri*). *r.* Thalamus. *gl.* Glands. *ec.* Tetradynamous stamens. *sti.* Stigma.——*Fig.* 27. Vertical section of the ovary of the Wallflower. *ov.* Ovules, each attached by a stalk to the placenta, *cn.* *cl.* Vertical spurious dissepiment called the replum.——*Fig.* 28. Fruit or siliqua of the Wallflower, showing the separation of its two valves from the replum.——*Fig.* 29. The silicula of Shepherd's Purse (*Capsella Bursa-pastoris*) in the act of dehiscing, showing the stalked pendulous seeds.——*Fig.* 30. Silicula of the Scurvy-grass (*Cochlearia officinalis*) in the act of dehiscing.——*Fig.* 31. The embryo of *Bunias orientalis.*

Division of the Order and Illustrative Genera.—This large and truly natural order has been divided into sub-orders according to the nature of the fruit, and also to the mode in which the embryo is folded. The latter is the only satisfactory arrangement.

The sub-orders founded on the nature of the fruit are as follows:—

Sub-order 1. *Siliquosæ.*—Fruit a siliqua opening by valves longitudinally (*fig.* 28). *Illustrative Genera:*—Cheiranthus, *Linn.*; Brassica, *Linn.*

FIG. 32.

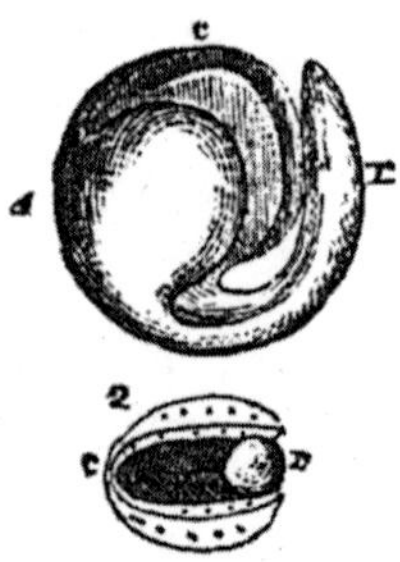

FIG. 33.

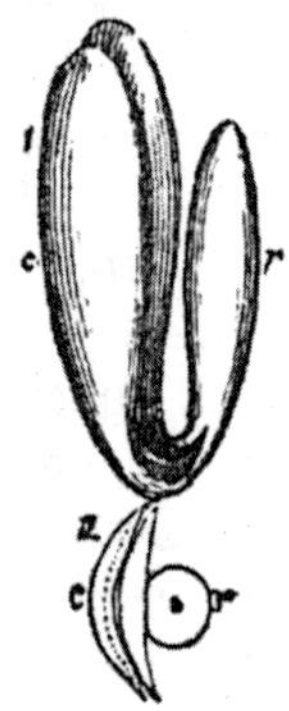

FIG. 34.

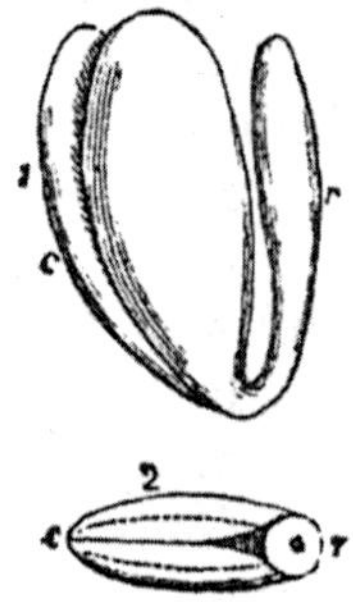

Fig. 32. The embryo of the Cabbage plant (*Brassica oleracea*). 1. Undivided. 2. Horizontal section. *r*. Radicle. *c*. Cotyledons. ——*Fig.* 33. Embryo of the Woad (*Isatis tinctoria*). 1. Undivided. 2. Horizontal section. *c*. Cotyledons. *r*. Radicle.——*Fig.* 34. Embryo of the Wallflower (*Cheiranthus Cheiri*). 1. Undivided. 2. Horizontal section. *r*. Radicle. *c*. Cotyledons.

Sub-order 2. *Siliculosæ latiseptæ.*—Fruit a silicula opening by valves; the replum in its broader diameter (*fig.* 30). *Illustrative Genus:*—Cochlearia, *Linn.*

Sub-order 3. *Siliculosæ angustiseptæ.*—Fruit a silicula opening by valves; the replum in its narrower diameter (*fig.* 29). *Illustrative Genera:*—Capsella, *Mœnch.*; Iberis, *Linn.*

Sub-order 4. *Nucumentaceæ.*—Fruit an indehiscent silicula; often 1-celled owing to the absence of the replum. *Illustrative Genus:*—Isatis, *Linn.*

Sub-order 5. *Septulatæ.*—The valves of the fruit opening longitudinally, and bearing transverse septa in their interior. No examples among British plants.

Sub-order 6. *Lomentaceæ.*—Fruit a siliqua or silicula, dividing transversely into 1-seeded portions, the true siliqua sometimes barren; the beak placed above is containing one or two seeds. *Illustrative Genera:* —Cakile, *Gaert.*; Raphanus, *Linn.*

The sub-orders founded on the mode in which the embryo is folded are as follows:

Sub-order 1. *Pleurorhizeæ* (○ =) (*fig.* 34).—Cotyledons accumbent, flat; radicle lateral. *Illustrative Genera:*—Cheiranthus, *Linn.*; Arabis, *Linn.*

Sub-order 2. *Notorhizeæ* (○ ||) (*fig.* 33).—Cotyledons incumbent, flat; radicle dorsal. *Illustrative Genera:*—Hesperis, *Linn.*; Isatis, *Linn.*

Sub-order 3. *Orthoploceæ* (○ ≫) (*fig.* 32).—Cotyledons conduplicate, longitudinally folded in the middle; radicle dorsal, within the fold. *Illustrative Genera:*—Brassica, *Linn.*; Raphanus, *Linn.*

Sub-order 4. *Spirolobeæ* (○ || ||) (*fig.* 31).—Cotyledons twice folded, linear, incumbent. *Illustrative Genus:*—Bunias, *Linn.* There are no examples among British plants.

Sub-order 5. *Diplecolobeæ* (○ || || ||).—Cotyledons thrice folded, linear, incumbent. *Illustrative Genera:*—Senebiera, *DC.*; Subularia, *Linn.*

CARYOPHYLLACEÆ, the Pink Order.—Character. —*Herbs.* *Stems* swollen at the nodes. *Leaves* opposite, entire, often connate at their base; exstipulate, or rarely with small membranous stipules. *Inflorescence* cymose (*fig.* 36). *Flowers* generally hermaphrodite (*fig.* 40), or rarely unisexual. *Sepals* 4 or 5 (*figs.* 35 and 37), distinct (*fig.* 35), or united (*fig.* 37), persistent (*fig.* 45). *Petals* equal in number to the sepals (*fig.* 35), unguiculate (*fig.* 39), often deeply divided (*fig.* 39), sometimes absent, frequently raised above the calyx on a stalk (*fig.* 38). *Stamens* equal in number to the sepals, and then opposite to them, or usually twice as numerous (*figs.* 35 and 40), or rarely fewer, frequently attached with the petals on a stalk above the calyx (*fig.* 38); *filaments* generally distinct (*fig.* 40), or sometimes united at the base, subulate; *anthers* innate. *Ovary* sessile (*fig.* 40), or supported with the

FIG. 35.

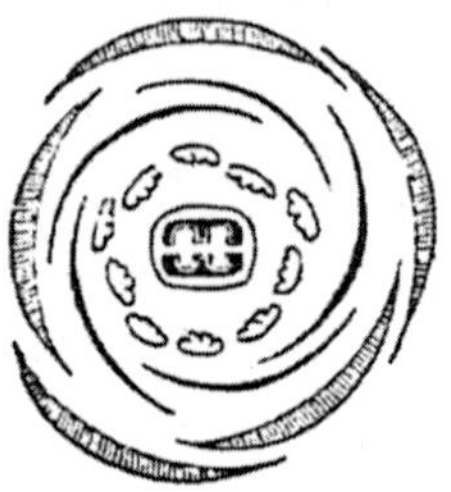

FIG. 36.

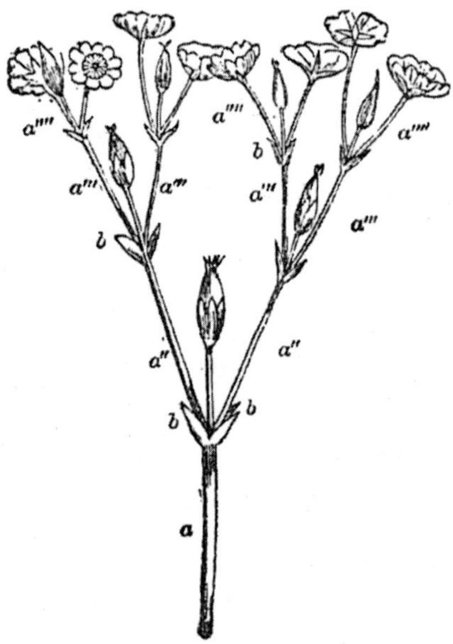

FIG. 37.

Fig. 35. Diagram of the flower of a species of *Dianthus.*——*Fig.* 36. Dichasial cyme of a species of *Cerastium.*——*Fig.* 37. Dentate or toothed calyx of a species of Campion (*Lychnis*).

petals and stamens on a short gynophore (*figs.* 41, *g*, and 38), 1-celled generally with a free central placenta (*fig.* 42, *p*), or rarely 2—5-celled (*figs.* 35 and 43); *styles* 2 (*fig.* 41) to 5 (*fig.* 42), papillose on their inner surface (*fig.* 41), and hence should be properly regarded as stigmas; *ovules* few or numerous (*fig.* 42, *g*), amphitropal. *Fruit* a 1-celled capsule, opening by 2—5 valves, or by 4—10 teeth at the apex (*figs.* 44 and

45), and having a free central placenta, or rarely 2—5-celled with a loculicidal dehiscence, and with the placentas slightly attached to the dissepiments. *Seeds* generally numerous, rarely few; *embryo* usually curved round mealy albumen (*figs.* 46 and 47, *emb*), or rarely straight.

Diagnosis.—Herbaceous plants with stems swollen at the nodes, and opposite entire exstipulate leaves; or rarely with small membranous stipules. Inflorescence

FIG. 38.

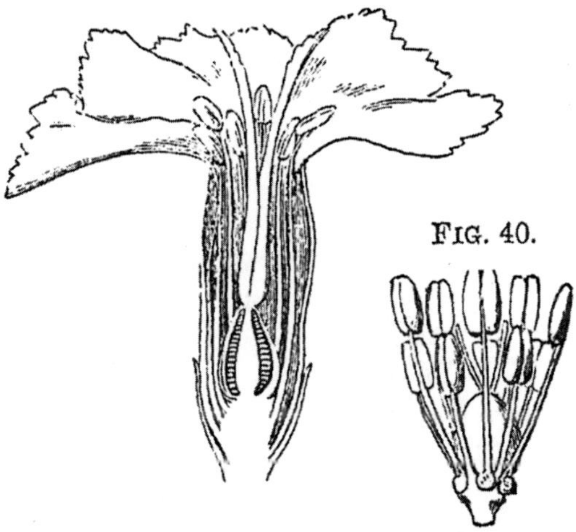

FIG. 39.

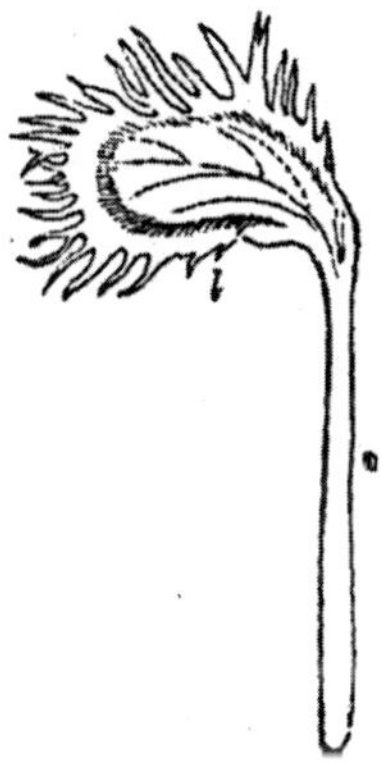

FIG. 40.

Fig. 38. Vertical section of the flower of a species of *Dianthus.*——*Fig.* 39. One of the petals of a species of *Dianthus.* *o.* Claw or unguis. *l.* Limb, which is fimbriated or fringed at the margins. ——*Fig.* 40. Essential organs of a species of *Stellaria.*

cymose. Flowers usually hermaphrodite. Sepals, petals, and stamens with a quaternary or quinary arrangement, the petals sometimes absent. Calyx persistent. Stamens hypogynous; anthers innate. Ovary superior, commonly 1-celled, styles 2—5. Capsule 1-celled, or rarely 2—5-celled; placenta usually free central, or in the 2—5-celled fruit slightly attached to the dissepiments. Seeds with the embryo usually curved round mealy albumen; or rarely straight.

Division of the Order and Illustrative Genera.—The order has been divided into three tribes as follows:—

FIG. 41.

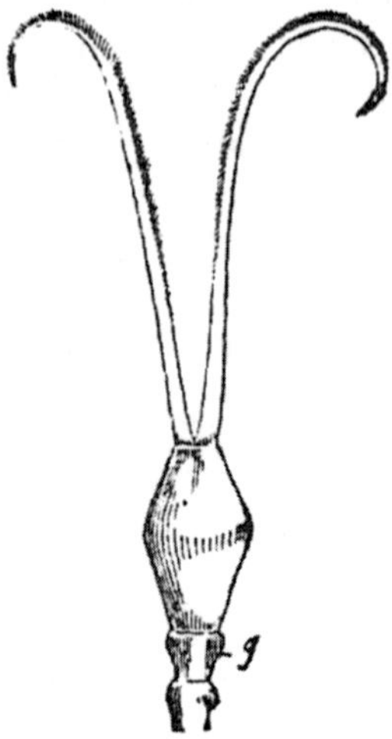

FIG. 42.

FIG. 43.

FIG. 46.

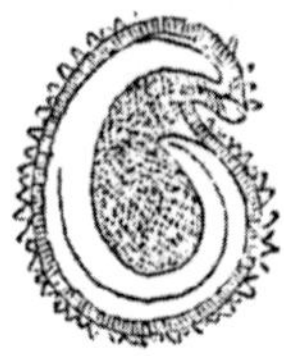

FIG. 45.

FIG. 44.

FIG. 47.

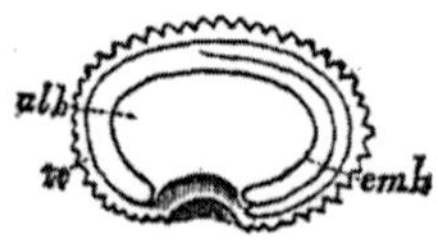

Fig. 41. Pistil of *Dianthus Caryophyllus* on a stalk, *g*, called the gynophore. On the top of the ovary are two styles, the face of each of which is traversed by a continuous stigmatic surface.——*Fig.* 42. Vertical section of the pistil of *Cerastium hirsutum*. *o*. Ovary. *p*. Free central placenta. *g*. Ovules. *s*. Styles and stigmas.——FIG. 43. Transverse section of the young ovary of a species of Campion (*Lychnis*) showing five partitions proceeding from the walls of the ovary to the placentas in the centre; these partitions are destroyed by the growth of the ovary, so that the placentation is ultimately free central.——*Fig.* 44. Capsule of a species of *Dianthus*, dehiscing partially in a valvular manner so as to form four teeth at the apex.——*Fig.* 45. Fruit of a species of *Lychnis* surrounded by the persistent calyx, and dehiscing partially in a valvular manner so as to form ten teeth at the apex.——*Fig.* 46. Vertical section of the seed of Chickweed (*Stellaria media*).——*Fig.* 47. Vertical section of the seed of *Lychnis dioica*. *te*. Integuments. *emb*. Embryo on the outside of the albumen, *alb*.

Tribe 1. *Alsineæ.*—Sepals distinct. Ovary sessile (*fig.* 40). No stipules. *Illustrative Genera*:—Alsine, *Wahlenb.*; Stellaria, *Linn.*

Tribe 2. *Sileneæ.*—Sepals united (*fig.* 37). Ovary usually stalked (*fig.* 41, *g*). No stipules. *Illustrative Genera*:—Dianthus, *Linn.*; Lychnis, *Linn.*

Tribe 3. *Polycarpeæ.*—Sepals distinct. Ovary sessile. Stipules membranous. *Illustrative Genus*:—Polycarpon, *Linn.*

MALVACEÆ, the Mallow Order.—Character.—*Herbs*, *shrubs*, or *trees*. *Leaves* alternate, more or less divided in a palmate manner, stipulate. *Flowers*

FIG. 48.

FIG. 49.

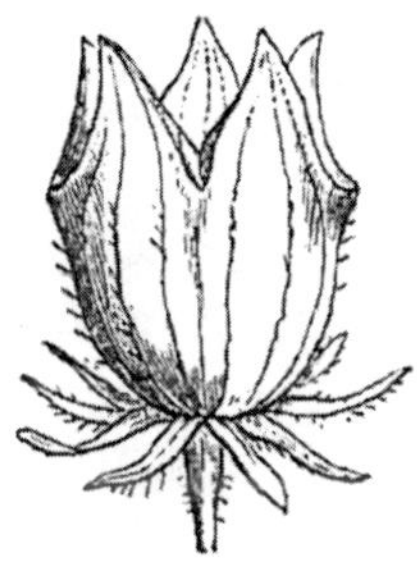

Fig. 48. Diagram of the flower of a species of *Malva*. The three external lines represent bracts, which together form an epicalyx or involucre.——*Fig.* 49. Calyx of *Hibiscus* surrounded by an epicalyx or involucre.

regular, usually axillary, and often surrounded by an involucre or epicalyx (*figs.* 48 and 49). *Sepals* usually 5 (*figs.* 48 and 49), rarely 3 or 4, more or less united (*figs.* 49 and 53); with valvate or some form of circular æstivation (*fig.* 48). *Petals* hypogynous, equal in number to the divisions of the calyx (*fig.* 48), with a twisted æstivation, either attached to the column formed by the united stamens (*fig.* 50) or free. *Stamens* hypogynous, numerous, monadelphous (*fig.* 51);

anthers 1-celled, reniform, with transverse dehiscence (*fig.* 52). *Ovary* superior (*fig.* 53), consisting of several

FIG. 50.

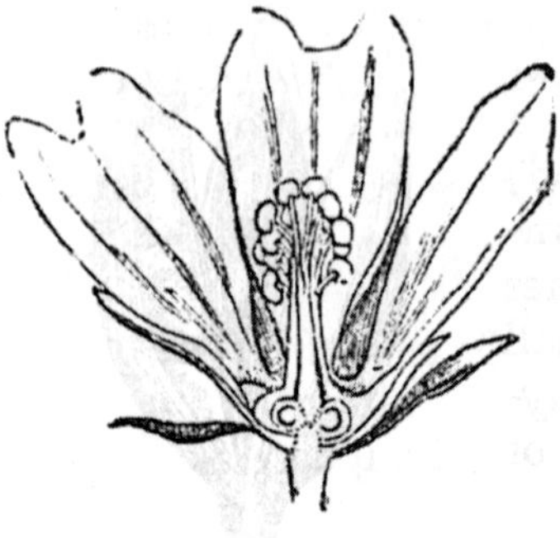

FIG. 51.

FIG. 52.

Fig. 50. Vertical section of the flower of a Mallow (*Malva*).——*Fig.* 51. Monadelphous stamens of a species of Mallow.——*Fig.* 52. Stamen of a Mallow, the anther of which has transverse dehiscence.

carpels (*figs.* 48 and 53), which are either apocarpous (*fig.* 53), or united so as to form a compound ovary

FIG. 53.

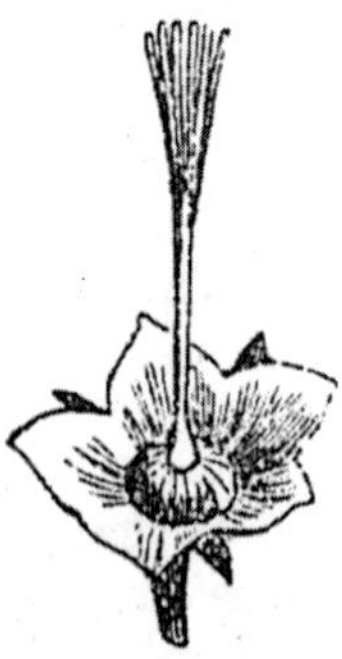

FIG. 54.

FIG. 55.

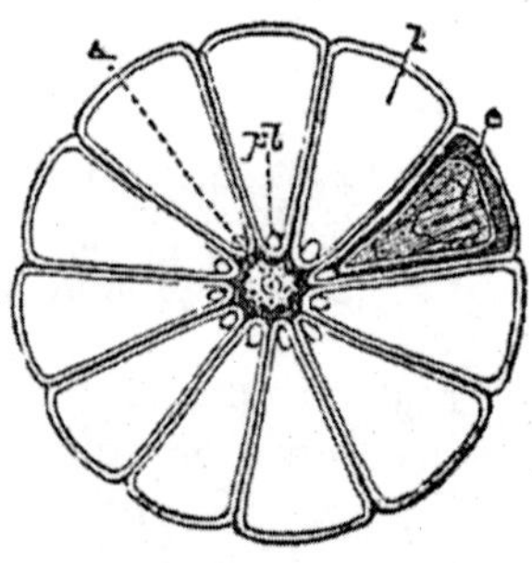

Fig. 53. Pistil of a Mallow surrounded by the inferior calyx and involucre.——*Fig.* 54. Carcerule or fruit of a Mallow, surrounded by the persistent calyx.——*Fig.* 55. Horizontal section of the fruit of *Malva sylvestris*. *a.* Axis. *pl.* Placenta. *l.* An empty cell. *c.* Embryo with twisted cotyledons.

with as many cells as there are carpels; *placentas* attached to the ventral sutures when the carpels are apocarpous (*fig.* 55, *pl*), or axile when the ovary is compound; *styles* equalling the carpels in number (*fig.* 53), united or distinct. *Fruit* either a carcerule, that is, consisting of a number of 1-celled, indehiscent (*figs.* 54 and 55), 1- or many-seeded carpels; or a capsule with loculicidal (*fig.* 56) or septicidal dehiscence, and numerous seeds. *Seeds* sometimes hairy; *albumen* none or in small quantity; *embryo* curved; *cotyledons* usually thin, folded or plaited (*fig.* 55, *c*).

FIG. 56.

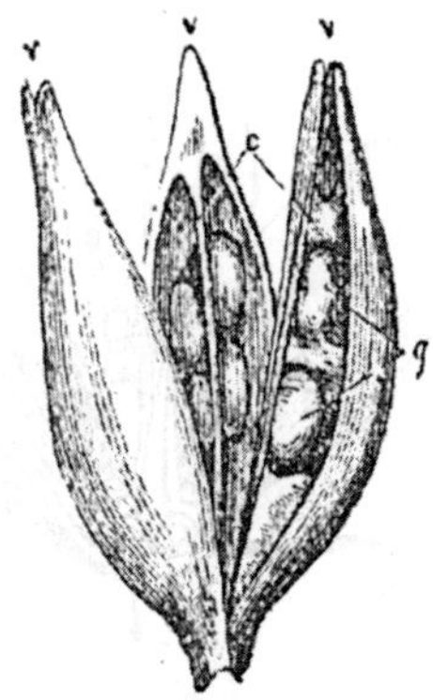

Fig. 56. Capsule of a species of *Hibiscus*, dehiscing loculicidally. *v*, *v*, *v*. Valves. c. Dissepiments. *g*. Seeds.

Diagnosis.—Leaves alternate, palmately-veined, simple, stipulate. Flowers regular. Calyx with valvate or some form of circular æstivation. Petals twisted in æstivation. Stamens hypogynous, numerous; anthers 1-celled, reniform, opening transversely; filaments united so as to form a column. Carpels distinct or united. Seeds with very little or no albumen; embryo curved; cotyledons folded or plaited.

Divisions of the Order and Illustrative Genera.—This order may be divided into three tribes as follows:—

Tribe 1. *Malveæ*—Flowers furnished with an involucre or epicalyx (*figs.* 48 and 53). Fruit consisting of separate carpels (apocarpous) (*figs.* 54 and 55). *Illustrative Genera*:—Malva, *Linn.*; Althæa, *Linn.*

Tribe 2. *Hibisceæ.*—Flowers furnished with an involucre (*fig.* 49). Fruit formed of united carpels (syncarpous) (*fig.* 56). *Illustrative Genera*:—Hibiscus, *Linn.*; Gossypium, *Linn.* No examples among British plants.

Tribe 3. *Sideæ.*—Flowers without an involucre. Fruit apocarpous or syncarpous. *Illustrative Genus*:—Sida, *Linn.* No examples among British plants.

POLYGALACEÆ, the Milkwort' Order.—Character.—*Shrubs* or *herbs*. *Leaves* alternate (*fig.* 57) or

FIG. 57.

FIG. 58.

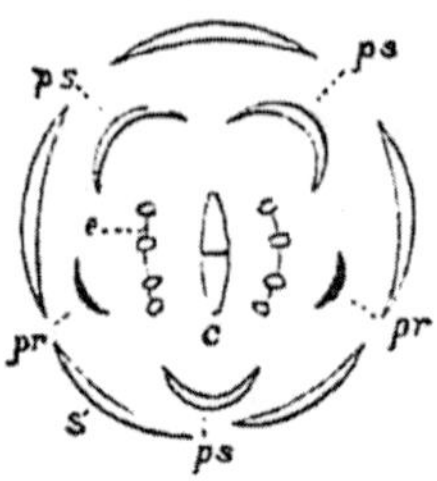

FIG. 59.

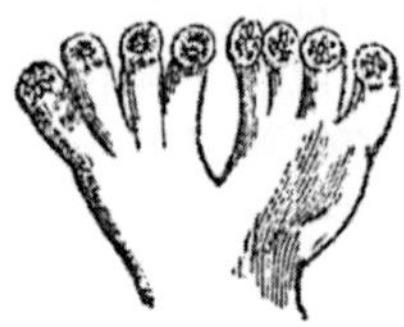

FIG. 60.

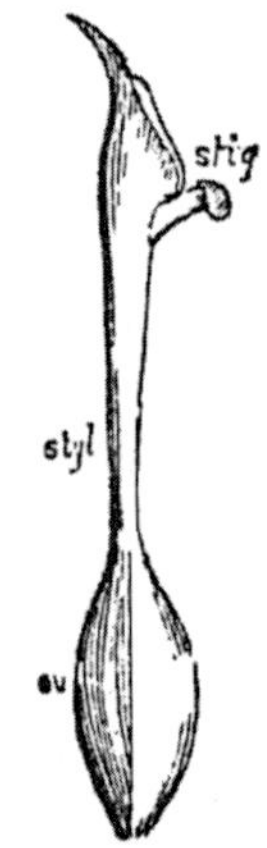

FIG. 61.

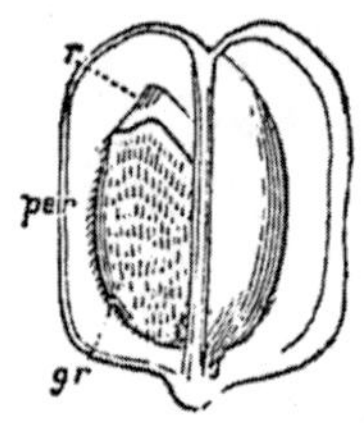

FIG. 62.

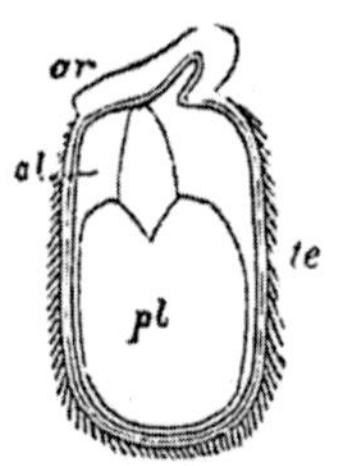

Fig. 57. A portion of the stem of the common Milkwort (*Polygala vulgaris*), with simple alternate exstipulate leaves, and irregular flowers.——*Fig.* 58. Diagram of the flower of the same. *s.* Sepals. *ps, ps, ps.* Posterior and anterior large petals. *pr, pr.* Lateral petals. *e.* Stamens. *c.* Carpels.——*Fig.* 59. Andrœcium of the same, with one-celled anthers dehiscing at their apex.——*Fig.* 60. Gynœcium of the same. *ov.* Ovary. *styl.* Style. *stig.* Stigma.——*Fig.* 61. Fruit with one cell opened. *per.* Pericarp. *gr.* Seed. *r.* Caruncule.——*Fig.* 62. Section of seed. *te.* Testa. *ar.* Caruncule. *al.* Albumen. *pl.* Embryo.

opposite, exstipulate, and usually simple (*fig.* 57). *Pedicels* bracteate. *Flowers* irregular, unsymmetrical (*figs.* 57 and 58), and somewhat papilionaceous in appearance; but here the *wings* belong to the calyx, whereas in the Leguminosæ they belong to the corolla. *Sepals* 5 (*fig.* 58, *s*), very irregular, usually distinct; of which 3 are placed exterior, and of these 1 is posterior and 2 are anterior; the 2 interior are lateral, usually petaloid (*fig.* 57), and form the wings to the flower. *Petals* hypogynous, usually 3, more or less united, of which 1, forming the keel, is larger than the rest, and placed at the anterior part of the flower; the keel is either naked, crested (*fig.* 57), or 3-lobed; the other 2 petals are posterior, and alternate with the wings and posterior sepal of the calyx, and are often united to the keel; sometimes there are 5 petals (*fig.* 58), and then the 2 additional ones, *pr*, *pr*, are of small size, and alternate with the wings and anterior sepals. *Stamens* hypogynous, 8 (*figs.* 58 *e*, and 59), usually combined into a tube, unequal, the tube split on the side next to the posterior sepal (*fig.* 59); *anthers* clavate, innate, usually 1-celled (*fig.* 59), or rarely 2-celled, opening by a pore at their apex, or rarely by valves. *Ovary* (*figs.* 58, *c*, and 60, *ov*) superior, 2—3-celled, one cell being frequently abortive; *ovules* solitary or twin, suspended; *style* simple (*fig.* 60, *styl*), curved, sometimes hooded at the apex; *stigma* simple (*fig.* 60, *stig*). *Fruit* (*fig.* 61) varying in its nature and texture, indehiscent or opening in a loculicidal manner, occasionally winged. *Seeds* pendulous (*fig.* 61, *gr*), smooth or hairy, with a caruncule next the hilum (*figs.* 61, *r*, and 62, *ar*); *embryo* straight or nearly so (*fig.* 62, *pl*), in copious fleshy albumen, *al*, and with the radicle towards the hilum.

Diagnosis.—Herbs or shrubs, with simple exstipulate leaves. Flowers irregular, unsymmetrical. Sepals and petals imbricate, not commonly corresponding in number, and usually arranged in a somewhat papilionaceous manner; odd petal anterior; odd sepal posterior. Stamens 8, hypogynous, usually combined; anthers generally 1-celled, with porous dehiscence. Fruit flattened, usually 2-celled and 2-seeded. Seed with abundant fleshy albumen, and with a caruncule next the hilum. *Illustrative Genus*:—Polygala, *Linn.*

LINACEÆ, the Flax Order. — Character.— *Herbs* or rarely *shrubs*. *Leaves* alternate or opposite, or rarely verticillate, entire, exstipulate. *Inflorescence* cymose. *Flowers* regular (*fig.* 63), symmetrical, generally very showy. *Calyx* imbricate, with 4 or 5

FIG. 63.

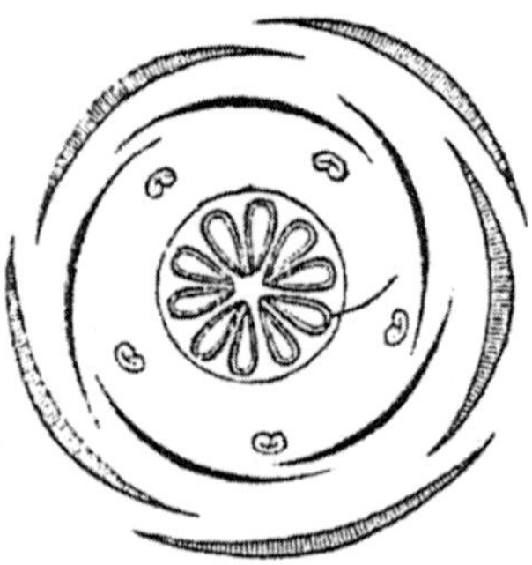

FIG. 64.

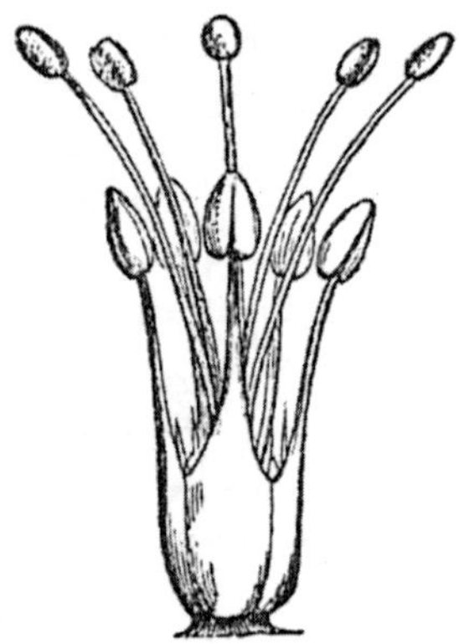

FIG. 65.

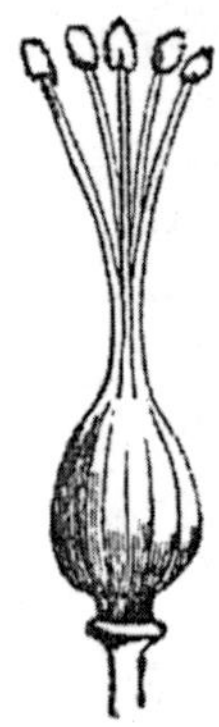

FIG. 66.

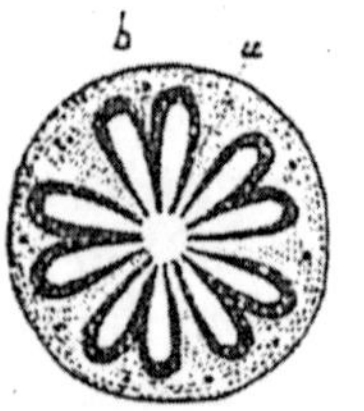

Fig. 63. Diagram of the flower of the Flax Plant (*Linum usitatissimum*).——*Fig.* 64. Essential organs of the same, showing the monadelphous stamens surrounding the pistil.——*Fig.* 65. Pistil of the same, with distinct styles and capitate stigmas.——*Fig.* 66. Transverse section of the ovary of the same, showing five complete and true dissepiments, *a*, and five incomplete spurious dissepiments, *b*.

sepals (*fig.* 63) persistent. *Petals* 4 or 5 (*fig.* 63), unguiculate, very deciduous, twisted in æstivation. *Stamens* 4—5, united at the base so as to form an hypogynous ring (*fig.* 64), from which proceed tooth-like processes (*staminodes*) alternating with the perfect stamens, and opposite to the petals. *Ovary* compound (*figs.* 63 and 66), superior, 3—5-celled; *styles* 3—5; *stigmas* capitate (*figs.* 64 and 65). *Fruit* capsular, many-celled, each cell more or less perfectly divided into two by a spurious dissepiment proceeding from the dorsal suture, and having a single seed in each division. *Seeds* compressed, with fleshy albumen, or rarely without albumen; *embryo* straight, with the radicle towards the hilum.

Diagnosis.—Herbs or very rarely shrubs, with exstipulate simple entire leaves. Inflorescence cymose. Flowers regular, symmetrical. Sepals, petals, and stamens 4—5 each; the sepals persistent and imbricate; the petals fugacious and twisted in æstivation; and the perfect stamens united at their base, and having little tooth-like staminodes alternating with them. Ovary 3—5-celled, styles distinct, stigmas capitate. Fruit capsular, each cell more or less divided by a spurious dissepiment, and each division containing one seed. Seeds compressed, with a straight embryo, and usually fleshy albumen. *Illustrative Genera*:—Linum, *Linn.*; Radiola, *Gmelin.*

GERANIACEÆ, the Crane's-bill Order.—Character.—*Herbs* or *shrubs*, with swollen joints (*nodes*). *Leaves* simple, opposite or alternate, stipulate. *Flowers* regular or irregular. *Sepals* 5, inferior, persistent (*figs.* 69 and 70, *c*), more or less unequal; *æstivation* imbricate. *Petals* 5 (*fig.* 67), or rarely 4 from abortion, unguiculate, hypogynous or perigynous; *æstivation* twisted (*fig.* 67). *Stamens* hypogynous, usually twice (*fig.* 68) or thrice as many as the petals; some are, however, frequently abortive, and generally united at the base, the alternate ones shorter (*fig.* 68), and occasionally barren. *Carpels* 5, arranged around an elongated axis or carpophore (*fig.* 69); *styles* corresponding in number to the carpels, and adhering to the carpophore. *Fruit* consisting of five one-seeded dehiscent carpels, which ultimately separate from the carpophore from below upwards by the curling up of the styles, which remain adherent at the summit (*fig.* 70). *Seeds* without albumen; *cotyledons* foliaceous, convolute (*fig.* 71).

Diagnosis.—Herbs or shrubs, with simple entire stipulate leaves, and swollen joints. Sepals 5, imbricate. Petals twisted in æstivation. Stamens generally united at the base, hypogynous. Fruit consisting of 5 ultimately dehiscent carpels attached by means of

FIG. 67.

FIG. 68.

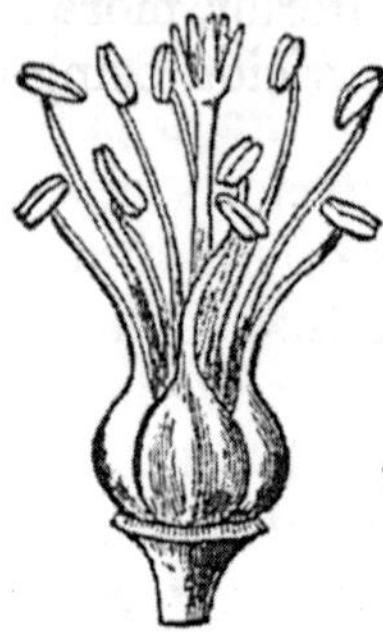

FIG. 69.

FIG. 70.

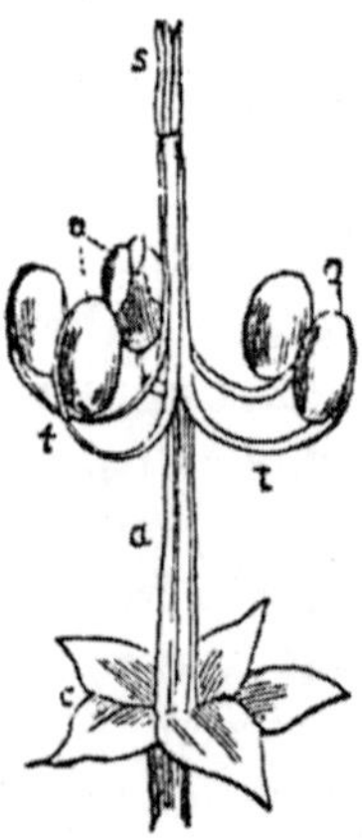

FIG. 71.

Fig. 67. A portion of the flowering stem of *Geranium sylvaticum.*——*Fig.* 68. The andrœcium and gynœcium of the same.——*Fig.* 69. The pistil partially matured, surrounded by the persistent calyx.——*Fig.* 70. Fruit of a species of *Geranium.* *c.* Persistent calyx. *a.* Axis or carpophore from which the mature ovaries, *o*, *o*, with their styles, *t*, *t*, are separating. *s.* Stigmas.——*Fig.* 71. Transverse section of the seed.

their styles to an elongated axis or carpophore, from which they separate when ripe from below upwards by the curling up of the styles. Seeds 1 in each carpel, exalbuminous; embryo convoluted. *Illustrative Genera* :—Erodium, *L'Héritier*; Geranium, *Linn.*; Pelargonium, *L'Héritier.*

Sub-class II. *Calyciflorœ.*

1. Perigynæ.

Leguminosæ, the Leguminous Order.—Character. —*Herbs, shrubs,* or *trees.*—*Leaves* alternate, stipulate, usually compound (*fig.* 72). *Calyx* (*figs.* 73, *s*, 74, *c*, and 75, *c*) monosepalous, inferior, more or less deeply divided into 5 parts, the odd division being anterior (*fig.* 73, *s*).

Fig. 72.

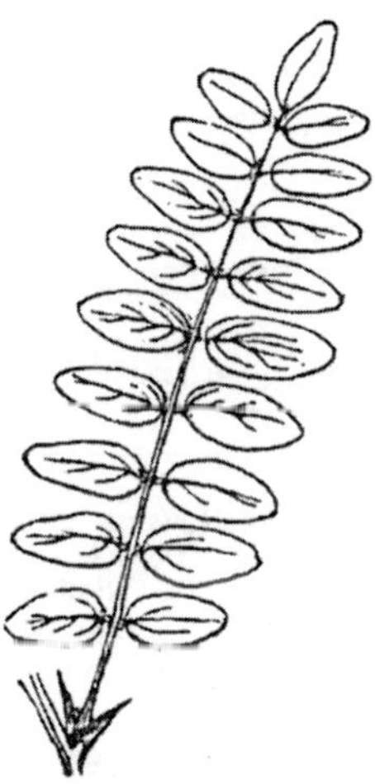

Fig. 73.

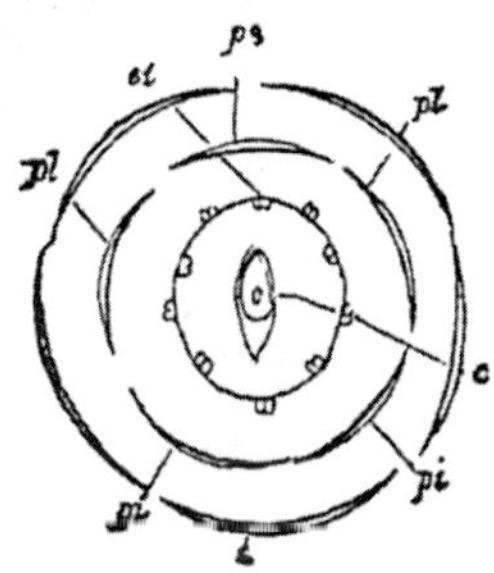

Fig. 74.

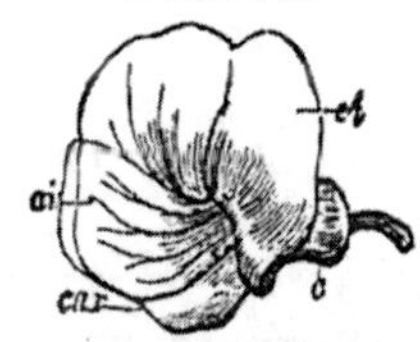

Fig. 72. Imparipinnate or unequally pinnate leaf of *Robinia*, with spiny stipules.——*Fig.* 73. Diagram of the flower of the Garden Pea (*Pisum sativum*). *s.* Anterior sepal. *ps.* Superior petal. *pi, pi.* Inferior petals. *pl, pl.* Lateral petals. *et.* Stamens. *c.* Carpel.——*Fig.* 74. The flower of the same. *et.* Standard or vexillum. *ai.* Wings or alæ. *car.* Carina or keel enclosing the essential organs. *c.* Calyx.

Petals usually 5 (*fig.* 73), or sometimes by abortion 4, 3, 2, 1, or rarely none, inserted into the base of the calyx, equal or unequal, often papilionaceous (*fig.* 74), the odd petal, if any, posterior (*figs.* 73, *ps*, 74, *et*). *Stamens* definite (*figs.* 75 and 76) or indefinite, usually perigynous, or rarely hypogynous, distinct or united into 1, 2 (*figs.* 75 and 76), or rarely 3, bundles. *Ovary* superior, usually formed of 1 carpel (*figs.* 73, *c*, and 77, *o*), although rarely of 2 or 5; 1-celled with 1, 2, or many ovules; *style* and *stigma* simple (*figs.* 75, *st*, and 77, *stig*). *Fruit* usually a legume (*fig.* 78), or sometimes a lomentum (*figs.* 79 and 80), or rarely a drupe. *Seeds* 1 or more, sometimes arillate, attached to the upper or ventral suture (*fig.* 81); *albumen* absent or present; *embryo* (*fig.* 82) straight or with the radicle folded upon the cotyledons; *cotyledons* leafy or fleshy, and either hypogeal or epigeal (*fig.* 83).

FIG. 75.

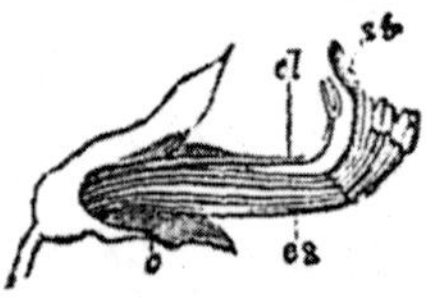

FIG. 76.

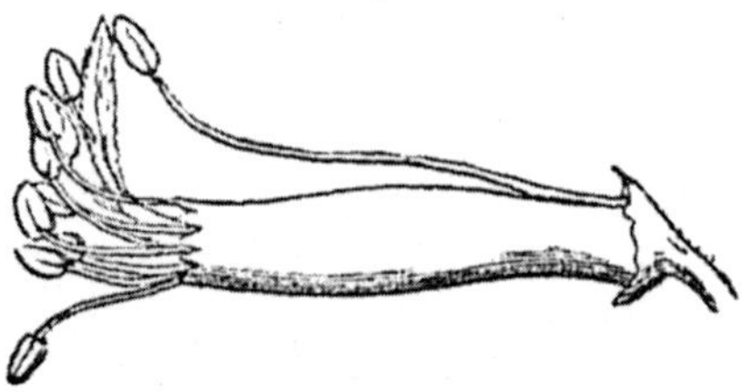

Fig. 75. The essential organs of the Garden Pea (*Pisum sativum*), surrounded by the calyx, *c*. *es*. Bundle of nine stamens. *el*. Solitary stamen. *st*. Style and stigma.——*Fig.* 76. Diadelphous stamens of the Sweet Pea (*Lathyrus odoratus*). There are ten stamens, nine of which are united and one distinct.

Diagnosis.—Herbs, shrubs, or trees. Leaves nearly always alternate and stipulate, and usually compound. Flowers regular or irregular. Calyx inferior, 5-partite; odd division anterior. Petals 5, and then often forming a papilionaceous corolla; or fewer by abortion, or none, perigynous; odd one, when present, posterior. Stamens distinct, or united into one or more bundles. Ovary superior, simple; style simple, proceeding from the ventral suture. Fruit usually a legume, or sometimes a lomentum, and rarely a drupe. Seeds 1 or more, with or without albumen. *This order may be usually distinguished by having papilionaceous corollas or leguminous fruit.*

Division of the Order and Illustrative Genera:—

The order has been divided into three sub-orders as follows:—

Sub-order 1. *Papilioneæ.*—Petals arranged so as to form a papilionaceous corolla, imbricate in æstivation,

FIG. 77.

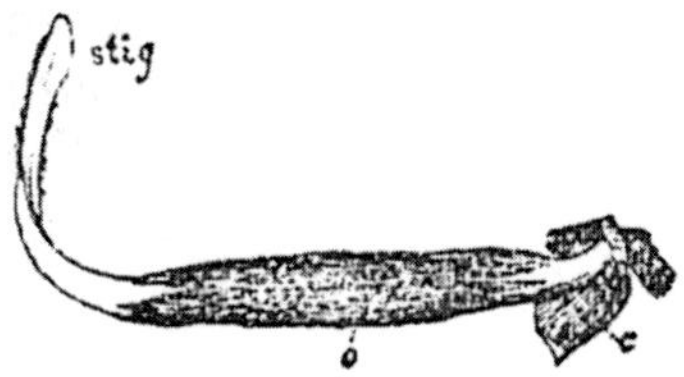

FIG. 78.

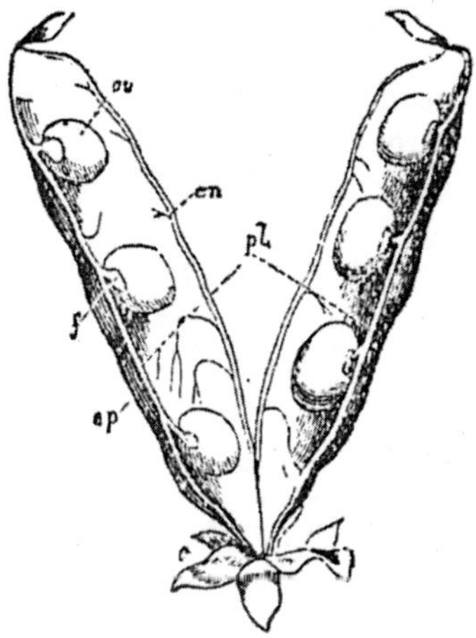

FIG. 79.

FIG. 80.

FIG. 81.

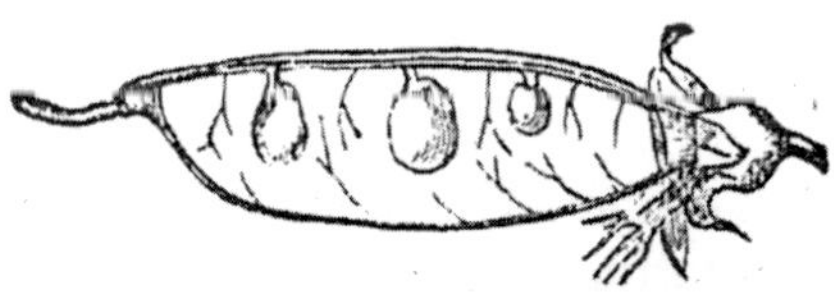

Fig. 77. Pistil of a species of *Lathyrus*. *o.* Ovary. *c.* Remains of the persistent calyx. On the top of the ovary is the simple style and stigma, *stig.*——*Fig.* 78. Legume of the Pea (*Pisum sativum*), which has opened by both dorsal and ventral sutures; hence it is two-valved. *c.* Persistent calyx. *ep.* Epicarp. *pl.* Placenta. *ov.* Seeds attached to the placenta by a funiculus or stalk, *f.* *en.* Endocarp.——*Fig.* 79. Lomentum of a species of *Acacia.*——*Fig.* 80. Lomentum of a species of *Hedysarum* separating transversely into one-seeded portions.——*Fig.* 81. Legume of the Garden Pea (*Pisum sativum*), with one valve removed.

and the upper or odd petal (*vexillum*) exterior (*fig.* 74, *et*). *Illustrative Genera*:—Ulex, *Linn.*; Trifolium, *Linn.*; Lotus, *Linn.*; Astragalus, *Linn.*; Vicia, *Linn.*; Ornithopus, *Linn.*

Sub-order 2. *Cæsalpinieæ.*—Petals not arranged in a papilionaceous manner, imbricate in æstivation, and the upper or odd petal inside the lateral petals. *Illustrative Genera* :—Cassia, *Linn.*; Tamarindus, *Linn.* There are no British plants in this sub-order.

FIG. 82.

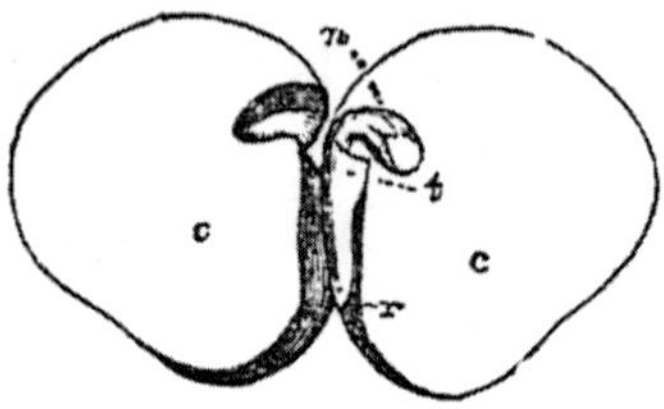

FIG. 83.

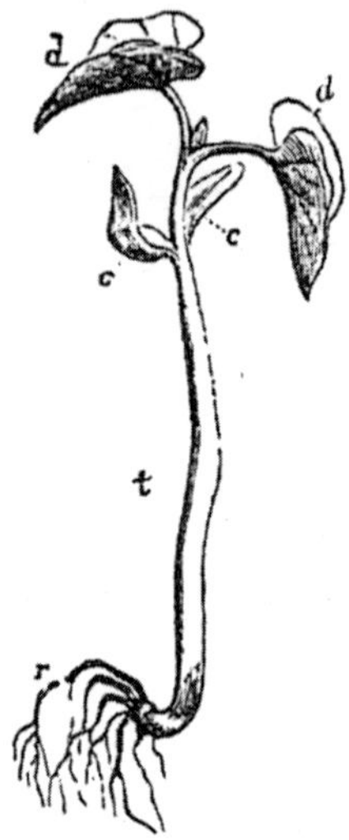

Fig. 82. Exalbuminous embryo of the Pea (*Pisum sativum*). *r.* The radicle. *t.* The axis (*tigellum*), terminated by the plumule, *n.* *c, c.* The two cotyledons.——*Fig.* 83. Germination of the Haricot or French Bean (*Phaseolus*). *r.* The roots, springing from the lower end of the axis, *t* (*tigellum*). *c, c.* The two cotyledons. *d, d.* The leaves.

Sub-order 3. *Mimoseæ.*—Petals equal, valvate in æstivation. *Illustrative Genera* :—Mimosa, *Linn.*; Acacia, *Willd.* There are no British plants in this sub-order.

ROSACEÆ, the Rose Order.—Character.—*Trees, shrubs,* or *herbs.*—*Leaves* simple (*fig.* 84) or compound (*fig.* 85), alternate (*fig.* 84), usually stipulate (*fig.* 85, *s, s*). *Flowers* regular (*figs.* 88 and 89), generally hermaphrodite (*figs.* 88 and 90), or rarely unisexual. *Thalamus* more or less convex (*figs.* 86 and 87, *l*), or concave (*fig.* 90). *Calyx* monosepalous (*figs.* 90—92), with a disk either lining the tube or surrounding the orifice, 4- or 5-lobed, when 5 the odd lobe posterior (*fig.* 88), sometimes surrounded by a whorl of bracts forming an involucre or epicalyx (*fig.* 91). *Petals* 5 (*figs.* 88 and 89, *p*), perigynous (*figs.* 90 and 94), rarely none (*fig.* 92).

FIG. 84.

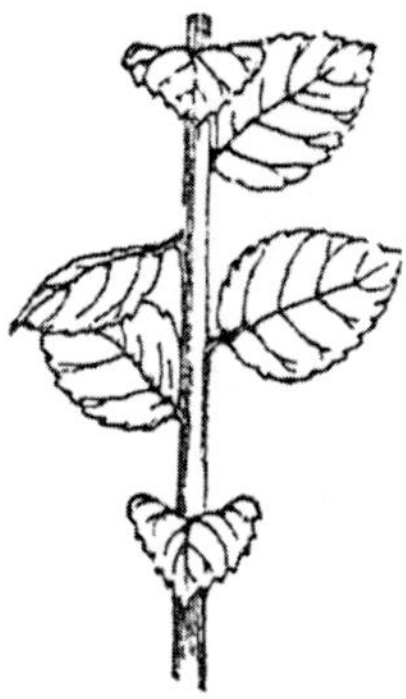

FIG. 85.

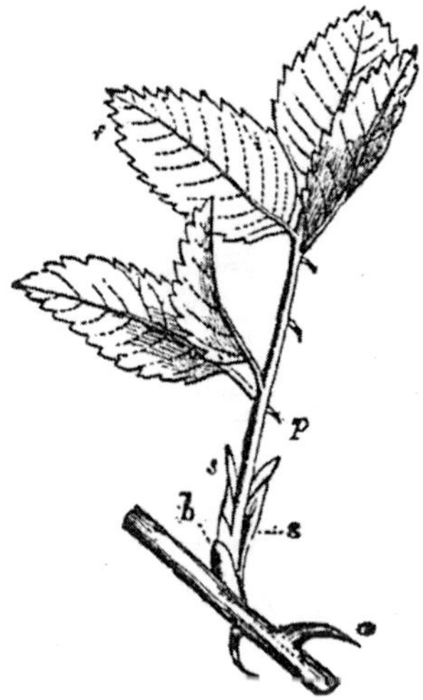

Fig. 84. A portion of a branch of the Cherry-tree (*Prunus Cerasus*), with the leaves alternately arranged.——*Fig.* 85. A portion of a branch of the common Rose (*Rosa canina*). *a.* A prickle. *b.* Bud in the axil of a compound leaf, with stalked leaflets. *s. s.* Adnate or adherent stipules.

FIG. 86.

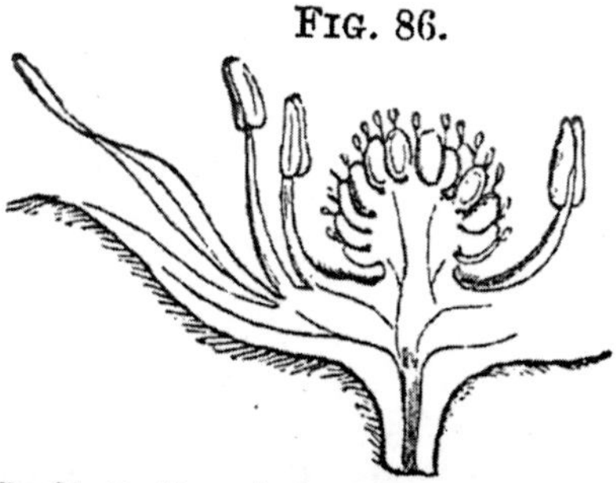

FIG. 87.

Fig. 86. Section of the flower of the Strawberry (*Fragaria*). The thalamus is nearly hemispherical, and bears a number of separate carpels on its upper portion.——*Fig.* 87. Section of the fruit of the Raspberry (*Rubus Idæus*), showing the conical thalamus, *l.*

Stamens definite (*fig.* 92), or numerous, perigynous (*figs.* 88—90); *anthers* (*fig.* 93) 2-celled, dehiscing longitudinally. *Carpels* 1 (*fig.* 94), 2, 5, or numerous (*figs.* 86 and 88), with 1-celled ovaries (*figs.* 92, 94,

FIG. 88.

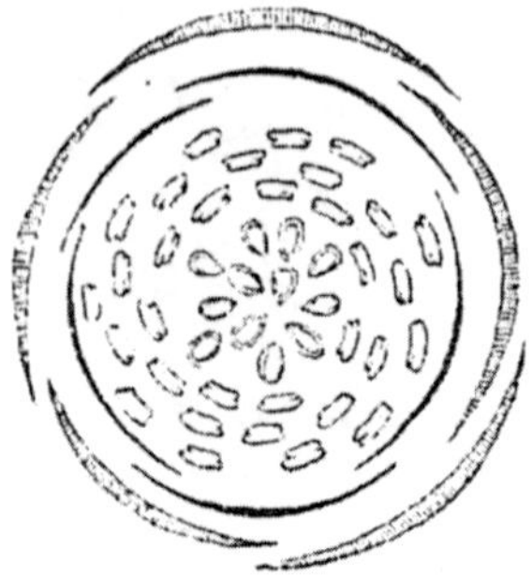

FIG. 89.

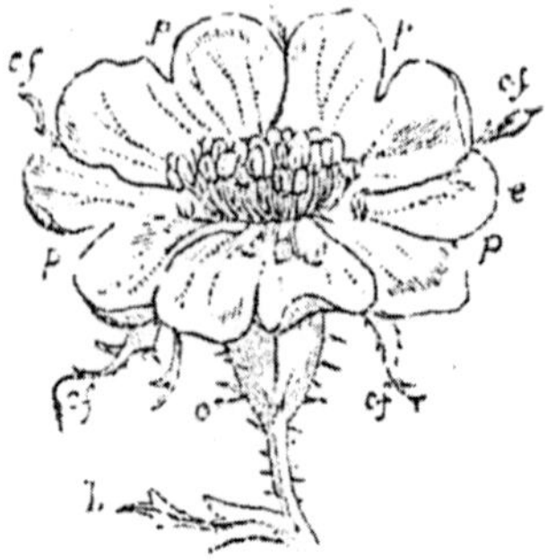

Fig. 88. Diagram of the flower of a species of Rose, with five sepals, five petals, numerous stamens, and many distinct carpels.—— *Fig.* 89. Flower of the Rose. *b.* Bract. *ct.* Tube of the calyx. *cf, cf, cf, cf, cf.* Divisions of the calyx. *p, p, p, p, p.* Petals.

FIG. 90.

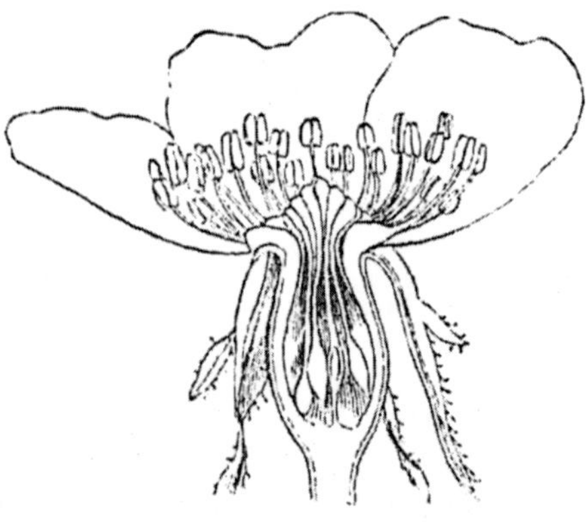

FIG. 91.

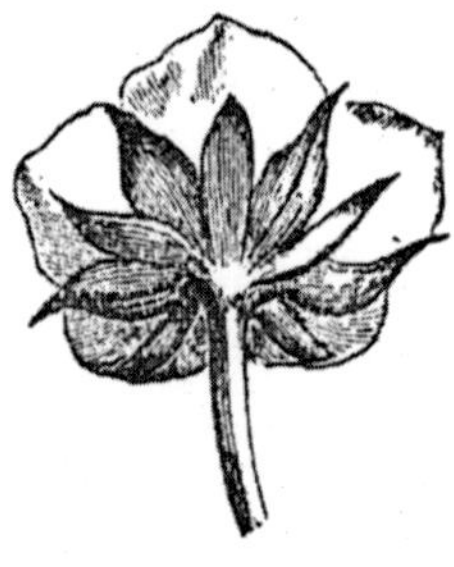

Fig. 90. Vertical section of the flower of a species of Rose.—— *Fig.* 91. Flower of the Strawberry (*Fragaria vesca*), showing the calyx, surrounded by an epicalyx or involucre.

and 96), usually apocarpous and superior (*fig.* 86), or sometimes more or less combined together, and with the tube of the calyx, and thus becoming inferior (*fig.* 95);

styles basilar (*figs.* 92 and 98), lateral (*fig.* 97), or terminal (*fig.* 94); *ovules* 1 (*figs.* 92 and 96), or few. *Fruit* various, either a drupe (*figs.* 99 and 100), an

FIG. 92.

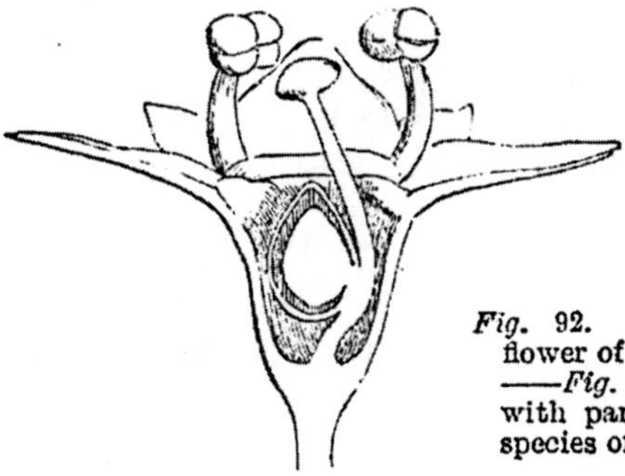

FIG. 93.

Fig. 92. Vertical section of a flower of a species of *Alchemilla.* ——*Fig.* 93. Two-celled anther with part of the filament of a species of *Rubus.*

achænium (*fig.* 101), a follicle, a dry or succulent etærio (*figs.* 102 and 103), a cynarrhodum, or a pome (*fig.* 104). *Seeds* 1 (*figs.* 100 and 101) or few (*fig.* 104), exalbuminous; *embryo* straight (*fig.* 100).

FIG. 94.

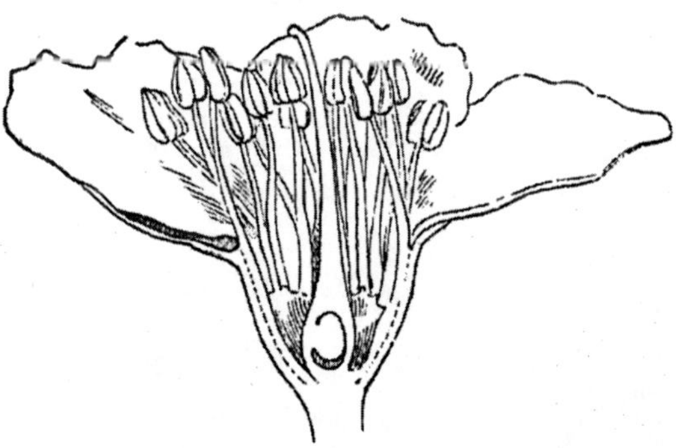

Fig. 94. Vertical section of the flower of the Peach (*Amygdalus persica*).

Diagnosis.—Trees, shrubs, or herbs, with alternate usually stipulate leaves. Flowers regular and generally hermaphrodite. Calyx 4—5-lobed; when 5, the odd lobe posterior. Petals 5, perigynous, or rarely none. Stamens perigynous, distinct; anthers

FIG. 95.

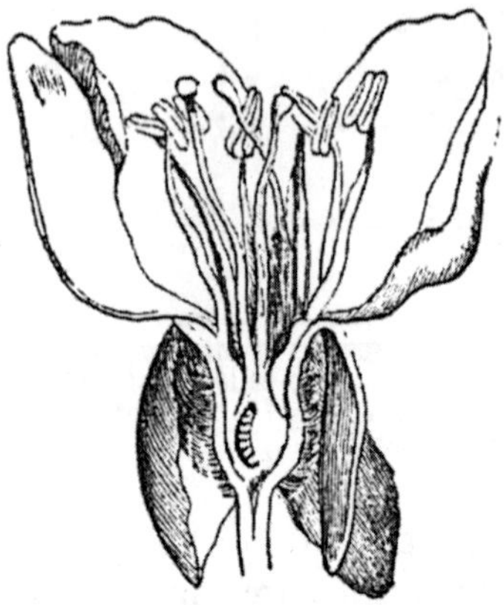

FIG. 96.

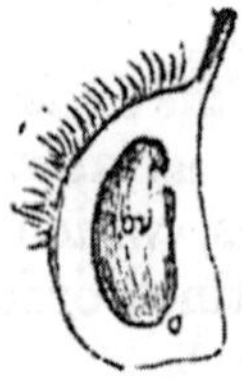

FIG. 97.

FIG. 98.

Fig. 95. Vertical section of the flower of the Quince (*Pyrus Cydonia*).——*Fig.* 96. Vertical section of the ovary, *o*, of a species of *Rubus*, with the ovule, *ov*.——*Fig.* 97. One of the carpels of the Strawberry with a lateral style.——*Fig.* 98. Carpel of *Alchemilla* with a basilar style. The stigma is capitate.

FIG. 99.

FIG. 100.

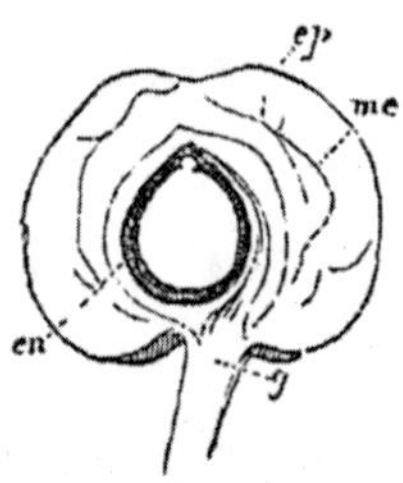

FIG. 101.

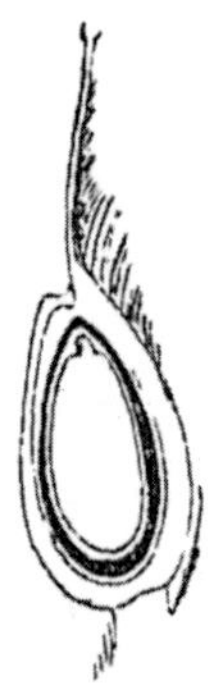

Fig. 99. Drupe of the Peach (*Amygdalus persica*).——*Fig.* 100. Vertical section of the fruit (*drupe*) of the Cherry (*Prunus Cerasus*). *ep.* Epicarp. *me.* Mesocarp. *en.* Endocarp, within which is the seed with embryo.——*Fig.* 101. Vertical section of an achænium of a species of Rose.

2-celled. Carpels one or more, usually distinct or sometimes united; generally superior or occasionally more or less inferior. Seeds 1 or few, exalbuminous; embryo straight.

Division of the Order and Illustrative Genera:—The order Rosaceæ, as above defined, may be divided into five sub-orders, which are by some botanists considered as distinct orders. They are characterised as follows:—

Sub-order 1. *Chrysobalaneæ.*—*Trees* or *shrubs*, with

FIG. 102.

FIG. 103.

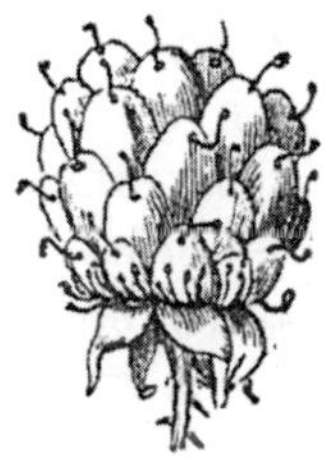

FIG. 104.

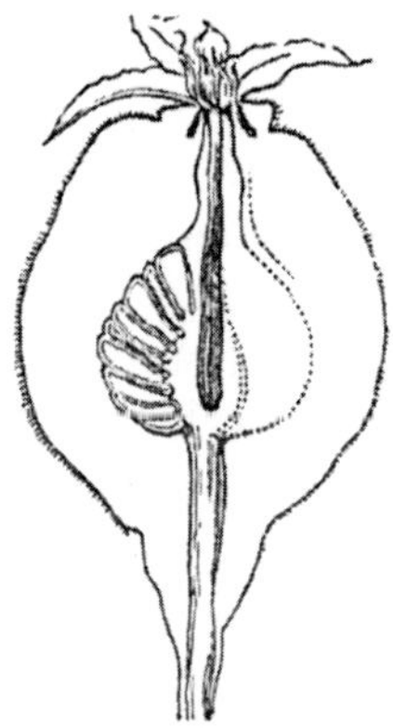

Fig. 102. Fruit (*etærio*) of the Strawberry (*Fragaria*).——*Fig.* 103. Fruit (*etærio*) of the Raspberry (*Rubus Idæus*).——*Fig.* 104. Vertical section of the fruit (*pome*) of the Quince (*Pyrus Cydonia*), showing the tube of the calyx adherent to the mature carpels, and forming a part of the pericarp; the free portion, or limb, being foliaceous.

simple leaves and free stipules. Carpel solitary, cohering more or less on one side with the calyx; ovules 2; style basilar. Fruit a drupe. Seed erect. *Illustrative Genus:*—Chrysobalanus, *Linn.* There are no British plants in this sub-order.

Sub-order 2. *Drupaceæ.*—*Trees* or *shrubs*, with simple leaves (*fig.* 84), and free stipules. Calyx deciduous. Carpel solitary, not adherent to the calyx; style terminal. Fruit a drupe (*fig.* 99). Seed suspended (*fig.* 100). *Illustrative Genus*:—Prunus, *Linn.*

Sub-order 3. *Roseæ.*—*Shrubs* or *herbs*, with simple or compound leaves and adherent stipules (*fig.* 85). Carpels 1 or more, superior, not united to the tube of the calyx, distinct or sometimes more or less coherent; styles lateral (*fig.* 97), or nearly terminal. Fruit either an etærio, cynarrhodum, or consisting of several follicles. Seed usually suspended (*figs.* 96 and 101), or rarely ascending. *Illustrative Genera*: —Rosa, *Linn.*; Rubus, *Linn.*

Sub-order 4. *Sanguisorbeæ.*—*Herbs* or *undershrubs.* Flowers often unisexual. Petals frequently absent (*fig.* 92). Carpels 1–3; style terminal or basilar (*fig.* 98). Fruit an achænium enclosed in the tube of the calyx, which is often indurated. Seed solitary, suspended or ascending. *Illustrative Genera*:—Alchemilla, *Linn.*; Poterium, *Linn.*

Sub-order 5. *Pomeæ.*—*Trees* or *shrubs*, with simple or compound leaves and free stipules. Carpels 1 to 5, adhering more or less to each other and to the sides of the calyx, and thus becoming inferior; styles terminal. Fruit a pome (*fig.* 104), 1—5-celled or rarely spuriously 10-celled. Seeds ascending. *Illustrative Genera*:—Pyrus, *Linn.*; Cratægus, *Linn.*

2. Epigynæ.

CUCURBITACEÆ, the Gourd or Cucumber Order. —Character.—*Herbs*, generally of a succulent nature, and either prostrate or climbing by means of tendrils. *Leaves* succulent, alternate, exstipulate, with a radiated venation (*fig.* 105), more or less scabrous. *Flowers* unisexual (*figs.* 106 and 107), monœcious or diœcious. *Calyx* monosepalous, 5-toothed (*fig.* 106), the limb sometimes obsolete, superior in the female flowers (*fig.* 106, *co*). *Corolla* usually monopetalous (*figs.* 106, *p*, and 107, *p*), 4—5-partite, or sometimes of distinct valvate petals, occasionally fringed, with evident reticulated veins, perigynous (*fig.* 107). *Barren flower:* — *Stamens* usually 5, epipetalous (*fig.* 107, *st*), either distinct, monadelphous, or more frequently triadelphous (*fig.* 107, *st*) in such a way that two of the bundles contain each 2 stamens, and the other but 1 stamen; rarely there are but 2 or 3 stamens; *anthers* 2-celled, usually long and sinuous (*fig.* 108, *l*), rarely straight. *Fertile flower:*—*Ovary* inferior (*fig.* 106, *co*), 1-celled, or

FIG. 105.

FIG. 106.

Fig. 105. Leaf of the Melon (*Cucumis Melo*). The venation is radiated or palmately-veined.——*Fig.* 106. Female or pistillate flower of the Cucumber (*Cucumis sativus*). *co.* Calyx adherent to the ovary, the limb is seen above, with five divisions. *p.* Corolla. *s.* Stigmas.

usually spuriously 3-celled from the projection inwards of the placentas; *placentas* parietal, usually 3; *ovules* indefinite or sometimes solitary; *style* short (*figs.* 106 and 109); *stigmas* thickened (*figs.* 106, *s*, and 109), papillose, lobed (*fig.* 109), or fringed. *Fruit* a pepo (*figs.* 110 and 111), or rarely a succulent berry. *Seeds* more or less flattened, usually with a leathery or horny testa, solitary or numerous; *embryo* flat, without albumen, and with leafy cotyledons.

(This order is sometimes placed amongst the Corolliforæ on account of its commonly monopetalous flowers, but its affinities are so essentially with the epigynous Calycifloræ, that we have placed it here in accordance with De Candolle's views, and those of most other botanists.)

Fig. 107.

Fig. 108.

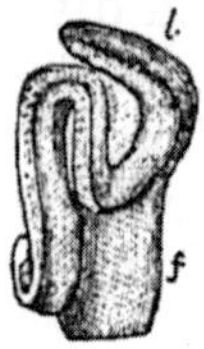

Fig. 111.

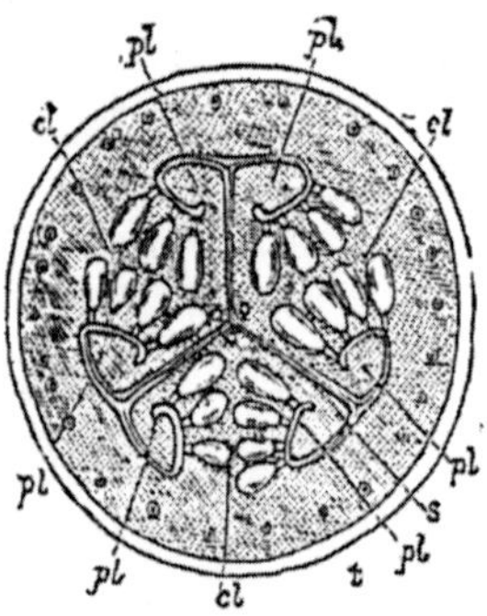

Fig. 109.

Fig. 110.

Fig. 107. Male or staminate flower of the Cucumber (*Cucumis sativus*), the floral envelopes of which have been divided in a longitudinal manner. *c.* Calyx. *p.* Corolla. *st.* Stamens.——*Fig.* 108. The sinuous anther, *l*, attached to the filament, *f*, of the common Bryony (*Bryonia dioica*).——*Fig.* 109. Lobed and thickened stigma of the Melon (*Cucumis Melo*).——*Fig.* 110. Pepo of the Squirting Cucumber (*Ecballium officinarum*), discharging its seeds.——*Fig.* 111. Transverse section of the fruit or pepo of the Melon. *cl, cl, cl.* Carpels. *pl, pl, pl, pl, pl, pl.* Curved placentas, sending processes, *s*, from the circumference, *t*, to the centre, and thus causing the fruit to be spuriously three-celled.

Diagnosis.—Herbs, usually of a succulent nature, prostrate or climbing. Leaves rough, alternate, radiated veined, exstipulate. Flowers unisexual. Calyx 5-toothed or obsolete, superior. Corolla perigynous. Barren flower with usually 5 stamens, which are distinct, monadelphous, or triadelphous, and epipetalous; rarely there are but 2 or 3 stamens; anthers long and usually sinuous. Fertile flower:—Ovary inferior, with parietal placentas; style short; stigmas more or less dilated. Fruit succulent. Seeds flat, exalbuminous, cotyledons leafy.

Division of the Order and Illustrative Genera.— This order has been divided by De Candolle into three sub-orders as follows:—

Sub-order 1.—*Nhandirobeæ.*—Anthers not sinuous. Placentas projecting so as to meet in the centre of the fruit. Seeds numerous. *Illustrative Genera:*— Telfairia, *Hook.*; Feuillæa, *Linn.* There are no examples among British plants.

Sub-order 2. *Cucurbiteæ.*—Anthers sinuous (*fig.* 108). Placentas projecting so as to meet in the centre of the fruit (*fig.* 111). Seeds numerous. *Illustrative Genus:*—Bryonia, *Linn. Bryonia dioica* is the only British plant in this order.

Sub-order 3. *Siceæ*—Placentas not projecting. Seed solitary, pendulous. *Illustrative Genera:*—Sicyos, *Linn.*; Sechium, *P. Br.* There are no examples among British plants.

The Cucurbitaceæ have been also divided by Bentham and Hooker into three series as follows:—

Series 1. *Plagiospermeæ.*—Ovules horizontal. *Illustrative Genus:*—Bryonia, *Linn.*

Series 2. *Orthospermeæ.*—Ovules erect or ascending. *Illustrative Genus:*—Trianosperma, *Torr. et Gr.* There are no examples among British plants.

Series 3. *Cremospermeæ.*—Ovules pendulous. *Illustrative Genera:*—Sicyos, *Linn.*; Sechium, *P. Br.* There are no examples among British plants.

ONAGRACEÆ, the Evening Primrose Order.— Character.—*Herbs* or *shrubs*. *Leaves* alternate or opposite, simple, exstipulate, without dots. *Calyx* superior (*fig.* 113), tubular, with the limb usually

4-lobed, or sometimes 2-lobed (*fig.* 112); in æstivation valvate. *Petals* usually large and showy, generally regular and equal in number to the divisions of the calyx (*fig.* 112), twisted in æstivation, and inserted into the throat of the calyx (*fig.* 113), rarely absent. *Stamens* (*figs.* 112 and 113) definite, 2, 4, or 8, or rarely by abortion 1, inserted with the petals into the throat of the calyx; *filaments* distinct; *pollen* trigonal (*fig.* 114). *Ovary* inferior (*fig.* 113), 2—4-celled; *pla-*

FIG. 112.

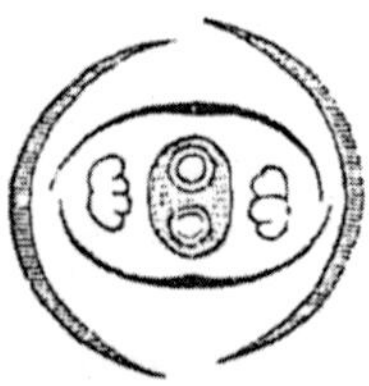

FIG. 113.

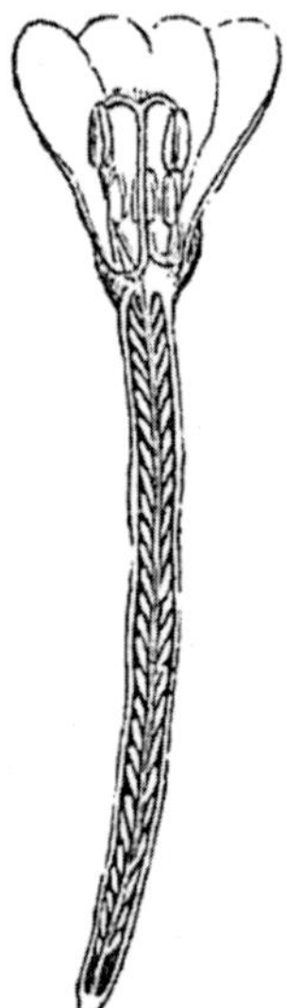

FIG. 114.

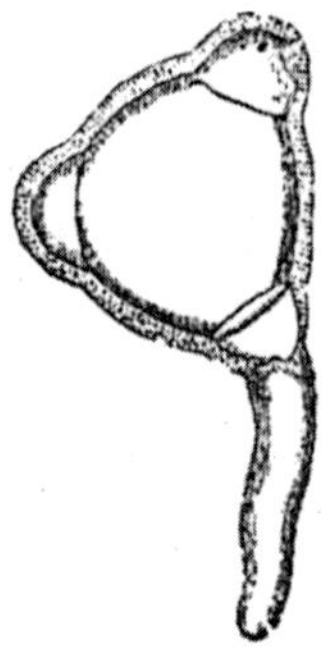

Fig. 112. Diagram of the flower of a species of *Circæa.*——*Fig.* 113. Vertical section of the flower of a species of Willow-herb (*Epilobium*).——*Fig.* 114. Trigonal pollen of a species of *Œnothera*, with a pollen-tube.

centas axile; *style* 1, filiform; *stigma* lobed or capitate. *Fruit* capsular, or succulent and indehiscent, 1, 2, or 4-celled. *Seeds* numerous, without albumen.

Diagnosis.—Herbs or shrubs, with simple exstipulate dotless leaves. Calyx superior, 2—4-lobed, valvate in æstivation. Petals usually equal in number to the lobes of the calyx, with a twisted æstivation, or rarely absent. Stamens few, inserted into the throat of the calyx with the petals. Ovary inferior, 2—4-celled; style simple; stigma lobed or capitate. Fruit dehiscent or indehiscent. Seeds numerous, without albumen. *Illustrative Genera*:—Œnothera, *Linn.*; Epilobium, *Linn.*; Circæa, *Tourn.*

UMBELLIFERÆ, the Umbelliferous Order.—Character.—*Herbs*, *shrubs*, or very rarely small *trees*, with usually hollow, or rarely solid, stems. *Leaves* alternate, generally amplexicaul (*fig.* 115), usually compound (*fig.* 115) or sometimes simple, and always exstipulate. *Flowers* generally in umbels, which are usually compound (*figs.* 116–118), or sometimes simple or capitate, with (*figs.* 117, *a*, and 118) or without (*fig.* 116, *a*) an involucre; the partial umbels or umbellules with (*figs.* 117, *b*, and 118, *b*) or without an involucel (*fig.* 116, *b*). *Calyx* (*fig.* 119) superior, the limb entire in the form of a ring, or 5-toothed, or obsolete.

FIG. 115.

Fig. 115. Amplexicaul petiole of *Angelica sylvestris*.

FIG. 116.

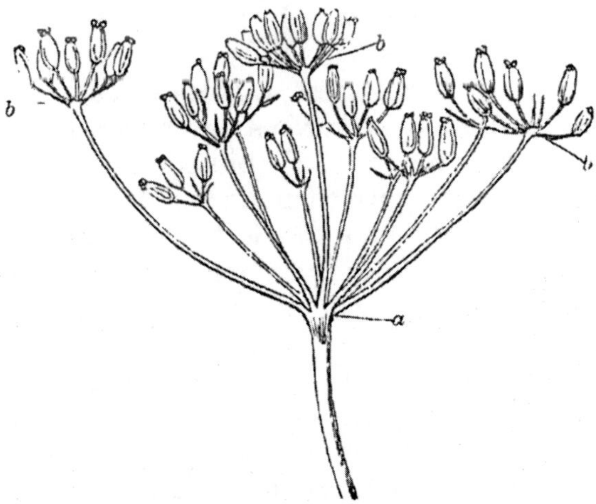

Fig. 116. Compound umbel of Fennel (*Fœniculum capillaceum*). *a.* General umbel. *b, b, b.* Partial umbels or umbellules.

Petals 5 (*fig.* 119), usually inflexed at the point, often unequal in size, inserted on the calyx outside the disk which crowns the ovary; *æstivation* imbricate, or rarely valvate or induplicate. *Stamens* 5, inserted with the petals and alternate with them (*fig.* 119), incurved in æstivation. *Ovary* inferior (*fig.* 119), crowned by a double fleshy disk (*stylopod*) *d*, 2-celled, with a solitary pendulous ovule in each cell; *styles* 2 (*fig.* 121, *g*); *stigmas* simple. *Fruit* called a *cremocarp* or *diachænium* (*figs.* 120 and 121), consisting of 2 carpels (*mericarps*), adhering by their face (*commissure*) to a common axis (*carpophore*), which is undivided (*fig.* 123, *i*) or forked (*fig.* 120), from which they ultimately separate and become pendulous (*fig.* 120); each mericarp (*figs.* 121 and 122) an indehiscent 1-seeded body (*fig.* 123), traversed on its dorsal surface by *ridges*, *a*, of which there are usually 5 (*fig.* 122), but sometimes there are 4 others, alternating with them, in which case the former are termed *primary*, and the latter *secondary* ridges; the spaces between the ridges are called channels (*valleculæ*), *b*, in which are sometimes linear oily receptacles called *vittæ* (*fig.* 124, 1, 2, 3, 4). *Seed* pendulous (*fig.* 123); *embryo* minute, *f*, at the base of abundant horny albumen, *d*; *radicle* pointing towards the hilum.

FIG. 117.

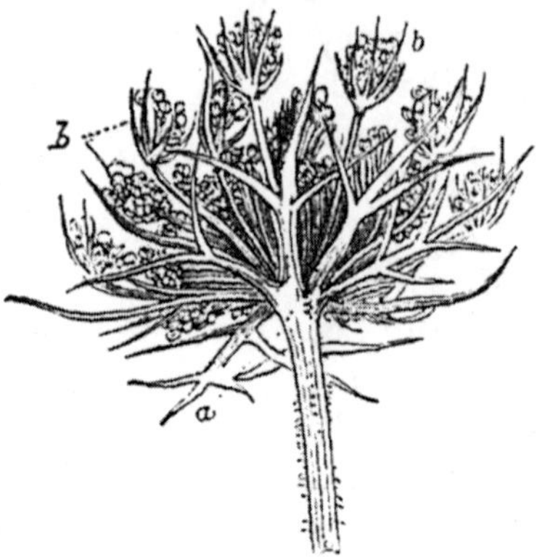

Fig. 117. Compound umbel of the Carrot (*Daucus Carota*). *a*. General involucre. *b*, *b*. Partial involucres or involucels.

Diagnosis.—Herbs or rarely shrubs. Stems generally hollow. Leaves alternate, usually compound and amplexicaul, or sometimes simple, and always exstipulate Flowers almost always arranged in a more or less umbellate manner. Calyx superior. Petals and stamens 5, inserted on the outside of a double fleshy epigynous disk. Ovary inferior, 2-celled, with a solitary pendulous ovule in each cell; styles 2. Fruit consisting of two indehiscent carpels, which separate,

FIG. 118.

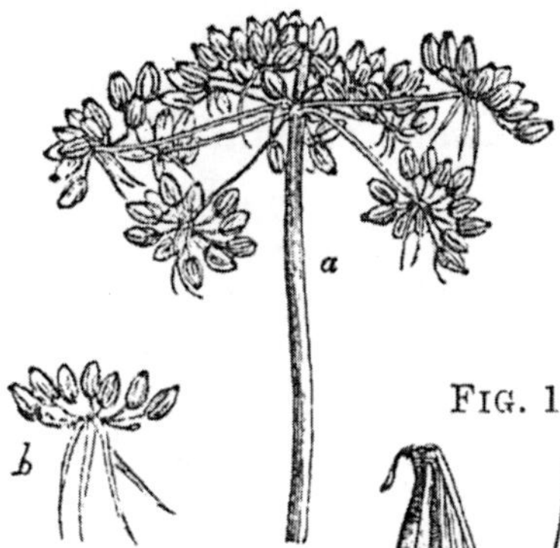

FIG. 119.

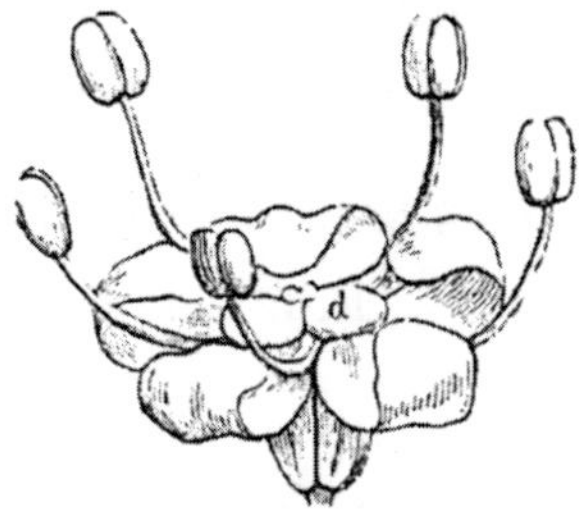

FIG. 120.

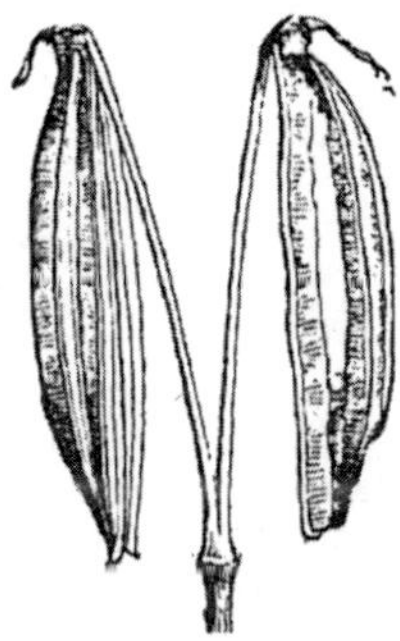

FIG. 121.

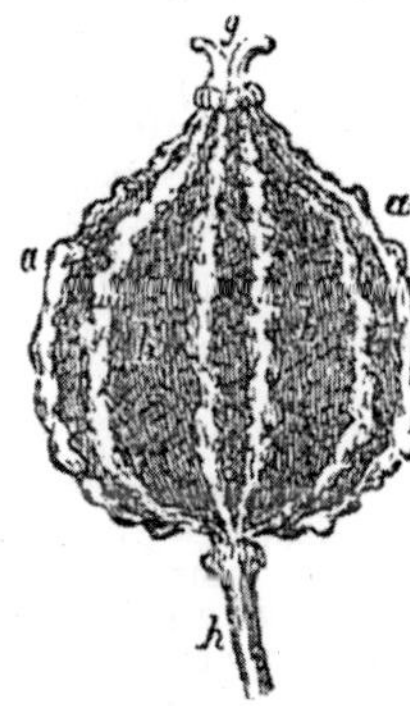

FIG. 122.

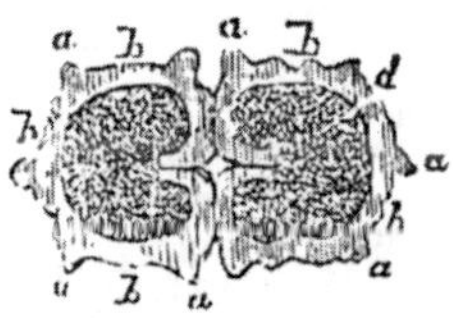

Fig. 118. *a.* General umbel of Fool's Parsley (*Æthusa Cynapium*) in fruit. *b.* One of the umbellules, showing the 3-leaved unilateral pendulous involucel.——*Fig.* 119. Flower of the Fennel (*Fœniculum capillaceum*). The ovary surmounted by a disk, *d.*——*Fig.* 120. Cremocarp or fruit of *Angelica sylvestris.*——*Fig.* 121. A side view of the ripe fruit of the Hemlock (*Conium maculatum*).——*Fig.* 122. Transverse section of the fruit of the same. (For explanation of letters, see *fig.* 122.)

when ripe, from a common axis or carpophore. Seeds pendulous, one in each carpel, with a minute embryo at the base of abundant horny albumen, and with the radicle towards the hilum.

Division of the Order and Illustrative Genera.—The order has been divided by De Candolle into three sub-orders from the appearance of the albumen, but these divisions are by no means well defined. They are as follows :—

Sub-order 1. *Orthospermeæ.*—Albumen flat on its face. *Illustrative Genera :*—Hydrocotyle, *Linn.*; Œnanthe, *Linn.*; Heracleum, *Linn.*

Sub-order 2. *Campylospermeæ.*—Albumen rolled inwards at the margins, and presenting a vertical furrow on its face. *Illustrative Genera :*—Anthriscus, *Hoffm.*; Chærophyllum, *Linn.*; Conium, *Linn.*

FIG. 123.

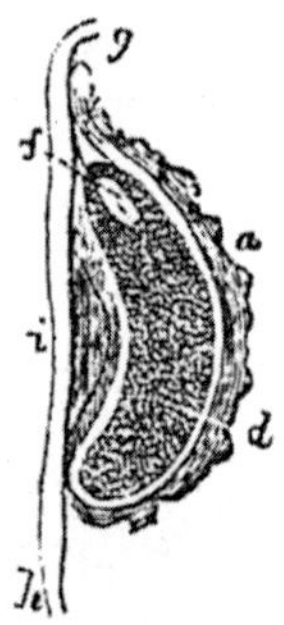

FIG. 124.

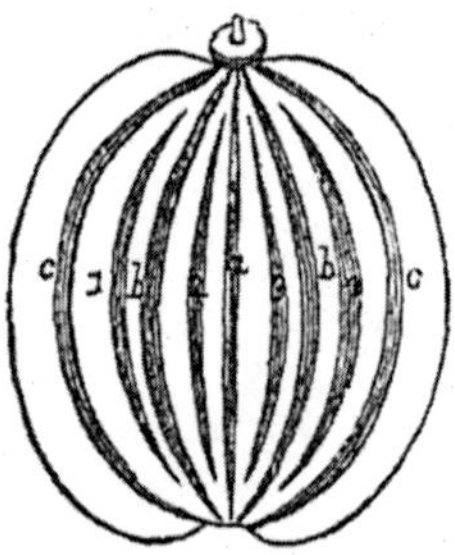

Fig. 123. Vertical section of one of the halves (*mericarps*) of the fruit of the Hemlock (*Conium maculatum*). The letters refer to the same parts in the three last figures. *a.* Ridges. *b.* Channels. *d.* Albumen. *f.* Embryo. *g.* Remains of the styles. *h.* Axis. *i.* Prolonged axis or carpophore.——*Fig.* 124. Dorsal surface of the fruit of the Parsnip (*Pastinaca sativa*). *a, b, b, c, c.* Primary ridges. 1, 2, 3, 4. Vittæ. The vittæ are readily seen by noticing that they are shorter than, and alternate with, the ridges.

Sub-order 3. *Cœlospermeæ.*—Albumen with the base and apex curved inwards on its face. *Illustrative Genus :*—Coriandrum, *Linn.*

This order has also been divided by Bentham and Hooker into three series as follows :—

Series 1. *Heterosciadieæ.*—Umbels generally simple or very irregularly compound, or the flowers are capitate. Vittæ none or obscure. *Illustrative Genera :*—Hydrocotyle, *Linn.*; Astrantia, *Linn.*; Eryngium, *Linn.*

Series 2. *Haplozygieæ.*—Umbels compound (*fig.* 116).

Primary ridges of fruit alone conspicuous (*figs.* 121 and 122). Vittæ usually obvious. *Illustrative Genera*:—Conium, *Linn.*; Myrrhis, *Scop.*; Fœniculum, *Adanson.*

Series 3. *Diplozygieæ.*—Umbels usually compound (*fig.* 117). Fruit with primary and secondary ridges generally well marked. *Illustrative Genera*:—Caucalis, *Linn.*; Daucus, *Linn.*

Sub-class III. *Corolliflore.*

1. Epigynæ.

RUBIACEÆ, the Madder Order.—Character.—*Trees*, *shrubs*, or *herbs*. *Stems* rounded (*fig.* 125, *r*), or angular (*fig.* 126). *Leaves* simple, entire, and either opposite (*fig.* 125, *f*, *f*) and with interpetiolar stipules, *s*, or whorled and exstipulate (*fig.* 126). (*Although practically we speak of whorled exstipulate leaves, the whorls are in reality partly formed of leaves and partly of*

FIG. 125.

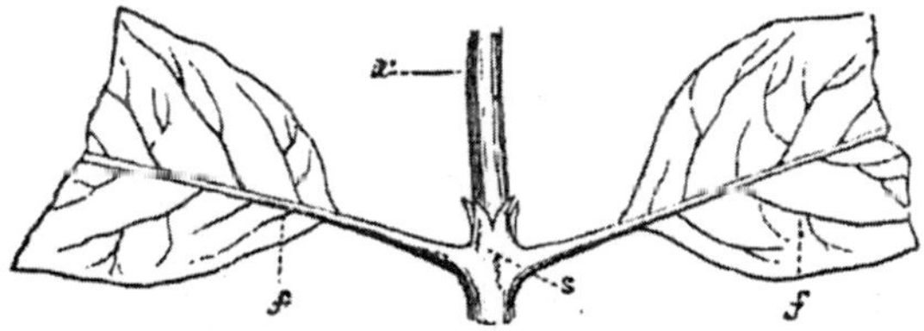

Fig. 125. A portion of a round branch, *r*, with two opposite leaves, *f*, *f*, of *Cephalanthus occidentalis*. *s*. Interpetiolar stipules.

stipules which resemble the true leaves in appearance.) *Inflorescence* cymose. *Calyx* superior (*figs.* 128, *cal*, and 129, *b*), with the limb 4—6-toothed, or entire (*fig.* 129, *b*), or obsolete (*fig.* 128). *Corolla* (*fig.* 127) epigynous, monopetalous, regular, tubular or rotate, with its lobes corresponding in number to the teeth of the calyx when the latter is divided. *Stamens* inserted upon the corolla and equal in number to, and alternate with, its lobes (*fig.* 127). *Ovary* inferior (*figs.* 128 and 129), crowned by a disk, usually 2-celled (*fig.* 127), or sometimes more; *ovules* anatropous; *style* 1 or 2 (*figs.* 128, *st*, and 129, *st*), (sometimes slightly divided; *stigma* simple (*figs.* 128 and 129) or divided. *Fruit* inferior,

FIG. 126.

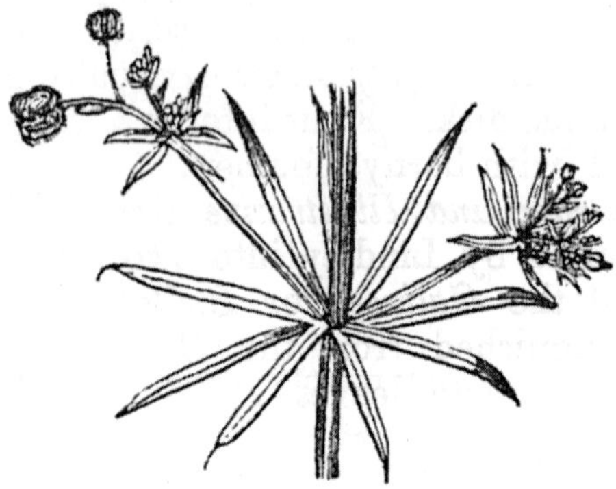

FIG. 127.

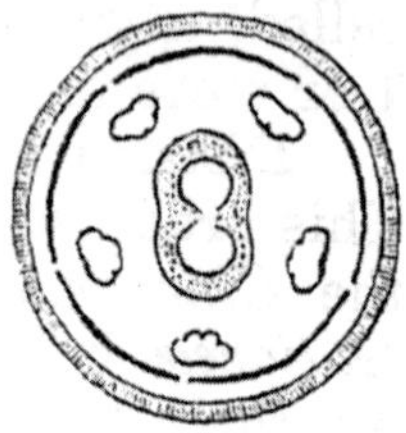

Fig. 126. Whorled leaves and angular stem of a species of *Galium.*——*Fig.* 127. Diagram of the flower of the Madder (*Rubia tinctorum*).

2-celled (*fig.* 130) or rarely more, dry or succulent, indehiscent or separating into two or more dry cocci. *Seeds* 1 (*fig.* 130), 2, or more, in each cell, when few

FIG. 128.

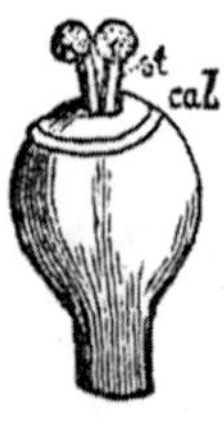

FIG. 129.

FIG. 130.

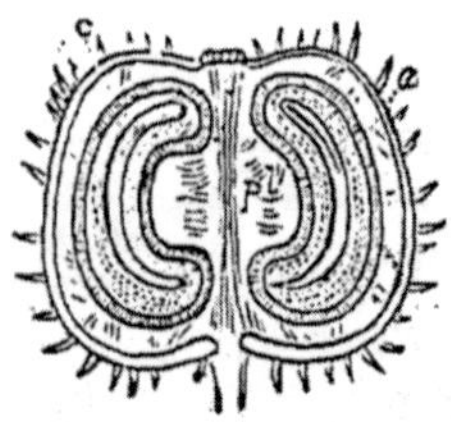

Fig. 128. Pistil of the Madder adherent to the calyx, *cal.* *st.* Styles and stigmas.——*Fig.* 129. Pistil of Goose Grass or Cleavers (*Galium Aparine*) surrounded by the calyx, *b.* *st.* Styles.——*Fig.* 130. Vertical section of the fruit and seeds of the same. *a.* Albumen. *c.* Embryo. *pl.* Placenta.

they are erect or ascending, or when numerous then attached to axile placentas; *embryo* small, in horny albumen (*fig.* 130, *a*).

Diagnosis.—Trees, shrubs, or herbs, with opposite

simple entire leaves, interpetiolar stipules, and rounded stems; or with whorled exstipulate leaves, and angular stems. Calyx superior. Corolla regular, epigynous. Stamens equal in number to the teeth of the calyx and segments of the corolla, with the latter of which they are alternate, epipetalous. Ovary inferior, usually 2-celled, with an epigynous disk. Fruit inferior. Seeds 1 or more in each cell, with horny albumen.

Division of the Order and Illustrative Genera:—This order was separated by Lindley into two orders, the *Cinchonaceæ* and the *Galiaceæ* or *Stellatæ*; the *Galiaceæ* being distinguished from the *Cinchonaceæ* by their whorled exstipulate leaves, angular stems, and other unimportant characters. But his views have not been adopted by botanists. The order *Rubiaceæ* is now divided by Bentham and Hooker into three series, each of which is again subdivided into sub-series and tribes, but they are too complicated to be described here. *Illustrative Genera*:—Galium, *Linn.*; Rubia, *Linn.*; Cinchona, *Linn.*; Ixora, *Linn.* The first two genera only contain British plants.

COMPOSITÆ, the Composite Order.—Character.—*Herbs* or *shrubs*. *Leaves* alternate or opposite, exstipulate. *Flowers* (florets) hermaphrodite (*figs.* 131–133), unisexual (*fig.* 134), or neuter, arranged in capitula (*figs.* 135–137), which are commonly surrounded by an involucre formed of a number of imbricated bracts (*phyllaries*) (*fig.* 136); the separate florets are also frequently furnished with membranous or scale-like bractlets (*paleæ*) (*fig.* 138, *b*, *b*). The *capitula* develop successively in a centrifugal order (*fig.* 137). *Calyx* superior (*figs.* 131–133, and 139 and 140), its limb either entirely abortive (*fig.* 139), or membranous (*fig.* 140), in which case it is entire or toothed, or it is pappose—that is, divided into bristles, or simple, or branched, or feathery hair-like processes (*fig.* 132, *a*). *Corolla* monopetalous (*figs.* 131–133, and 139 and 140), epigynous, tubular (*figs.* 139 and 140), ligulate (*figs.* 133 and 134), or bilabiate (*fig.* 131), 4—5-toothed (*figs.* 139 and 140), with a valvate æstivation. *Stamens* (*figs.* 131–133, *e*) 5, or rarely 4, inserted on the corolla, and alternate with its divisions; *filaments* distinct (*fig.* 141) or rarely monadelphous; *anthers* united into a tube (*syn-*

FIG. 131. FIG. 132. FIG. 133. FIG. 134.

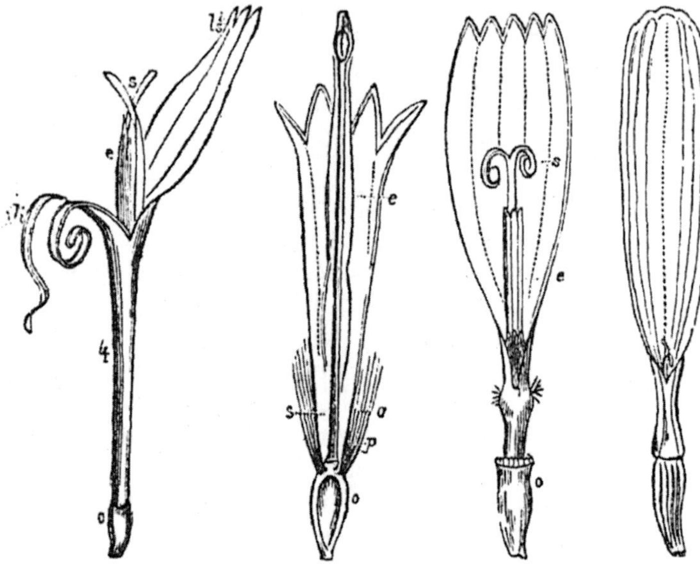

Fig. 131. Labiate floret of *Chælanthera linearis.* *o.* Ovary with adherent calyx. *t.* Tube of the corolla. *l s.* Upper lip of corolla. *l i.* Lower lip. *e.* Tube formed by the united anthers. *s.* Stigmas.——*Fig.* 132. Vertical section of the floret of *Aster rubricaulis.* *o.* Erect ovule, enclosed in the inferior ovary. *a.* Pappose limb of the calyx. *p.* Corolla. *s.* Style. *e.* Tube formed by the united anthers.——*Fig.* 133. Floret of the Chicory (*Cichorium Intybus*). *o.* Ovary with adherent calyx. *e.* Tube formed by the united anthers. *s.* Stigmas.——*Fig.* 134. Ligulate corolla of Ox-eye (*Chrysanthemum Leucanthemum*). The flower is unisexual (*pistillate*).

genesious or *synantherous*) (*fig.* 141), which is perforated by the style and stigmas (*figs.* 131, *s*, and 133, *s*). *Ovary* inferior, 1-celled (*fig.* 132), with 1 erect ovule; *style* 1, undivided below, and commonly bifid above (*fig.* 133); *stigmas* 2, one being usually placed on the inner surface of each division of the style (*fig.* 142,

1–8). *Fruit* a *cypsela*, dry, indehiscent, 1-celled, crowned by the limb of the calyx, which is often pappose (*fig.* 143). *Seed* (*fig.* 143) solitary, erect, exalbuminous; *radicle* inferior.

Diagnosis.—Herbs or shrubs, with exstipulate leaves. Flowers (called florets) arranged in dense capitula, which are commonly surrounded by an involucre. Capitula developed successively in a centrifugal order. Calyx superior, its limb abortive, or membranous, or pappose.

FIG. 135.

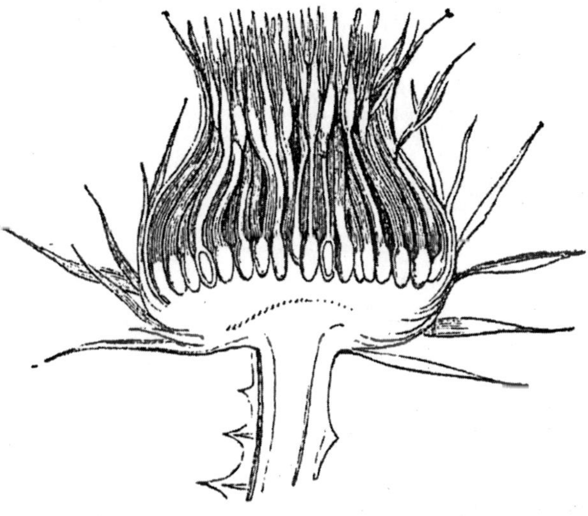

Fig. 135. Capitulum of the Cotton Thistle (*Onopordon Acanthium*).

Corolla epigynous, monopetalous, 4—5-toothed, with a valvate æstivation. Stamens epipetalous, equal in number to the divisions of the corolla (generally 5), and alternate with them; anthers syngenesious. Ovary inferior, 1-celled, with 1 erect ovule; style simple, bifid above; stigmas 2. Fruit 1-celled, dry, indehiscent. Seed solitary, erect, exalbuminous; radicle inferior.

Division of the Order and Illustrative Genera.—This order has been variously divided by botanists, but the arrangement most frequently adopted at the present day is as follows:—

SUB-ORDER 1. *Tubulifloræ.*—Florets all tubular and perfect; or those of the centre (*disk*) are tubular, and alone perfect, while those of the circumference (*ray*) are tubular or ligulate, and pistillate or neuter. Juice watery. It has been divided into five tribes:—

FIG. 136.

FIG. 137.

Fig. 136. Capitulum of the Marigold (*Calendula officinalis*), showing the florets enclosed in an involucre.——*Fig.* 137. Inflorescence of a species of *Senecio*, showing several capitula, which are seen to develop successively in a centrifugal order.

Tribe 1. *Vernonieæ.*—Style cylindrical; its arms generally long and subulate, sometimes short and blunt, always covered all over with bristles (*fig.* 142, 1). *Illustrative Genus:*—Vernonia, *Schreb.* There are no British plants in this tribe.

Tribe 2. *Eupatorieæ.*—Style cylindrical; its arms long and somewhat clavate, with a papillose surface on the outside near the end (*fig.* 142, 2). *Illustrative Genera* :—Eupatorium, *Tourn.*; Tussilago, *Tourn.*

Tribe 3. *Asteroideæ.*—Style cylindrical; its arms linear, flat on the outside, equally and finely downy on the inside (*fig.* 142, 3). *Illustrative Genera* :—Erigeron, *Linn.*; Bellis, *Linn.*

Tribe 4. *Senecioideæ.*—Style cylindrical; its arms linear, fringed at the point, generally truncate, but

FIG. 138.

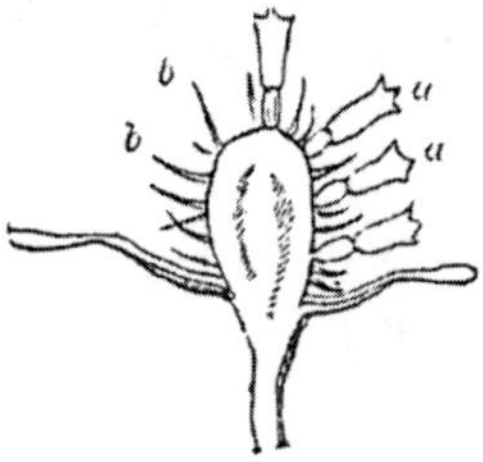

FIG. 139.

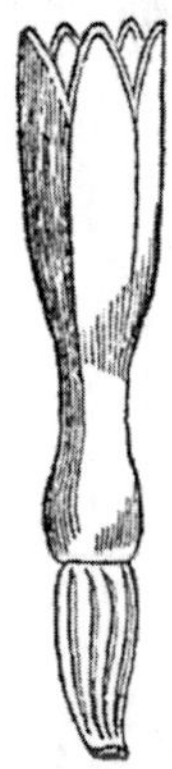

FIG. 140.

FIG. 141.

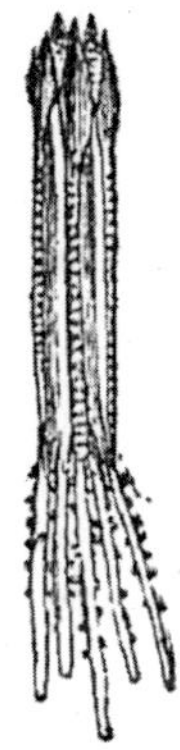

Fig. 138. Receptacle of the Chamomile (*Anthemis nobilis*) bearing flowers, *a*, *a*, and bracts or bracteoles, *b*, *b*: the latter are sometimes termed *paleæ*. The receptacle is here drawn much too large at the apex; it should be conical.——*Fig.* 139. One of the tubular florets of the Ox-eye (*Chrysanthemum Leucanthemum*). The calyx is completely united to the ovary, and presents no appearance of a limb.——*Fig.* 140. Tubular floret of the Sunflower (*Helianthus annuus*). The limb of the adherent calyx is membranous.——*Fig.* 141. Syngenesious anthers of a species of Thistle, the filaments being distinct.

sometimes extended beyond the fringe into a short cone or appendage of some kind (*fig.* 142, 4 and 5). *Illustrative Genera* :—Anthemis, *Linn.*; Senecio, *Linn.*

Tribe 5. *Cynareæ.*—Style thickened above, and often with a bunch or fringe of hairs at the enlarged portion; its branches united or free (*fig.* 142, 6). *Illustrative Genera* :—Arctium, *Linn.*; Centaurea, *Linn.*

SUB-ORDER 2. *Labiatifloræ.*—Florets with bilabiate corollas, perfect or unisexual. Juice watery. There are no British plants in this sub-order. Of this sub-order we have two tribes:—

FIG. 142.

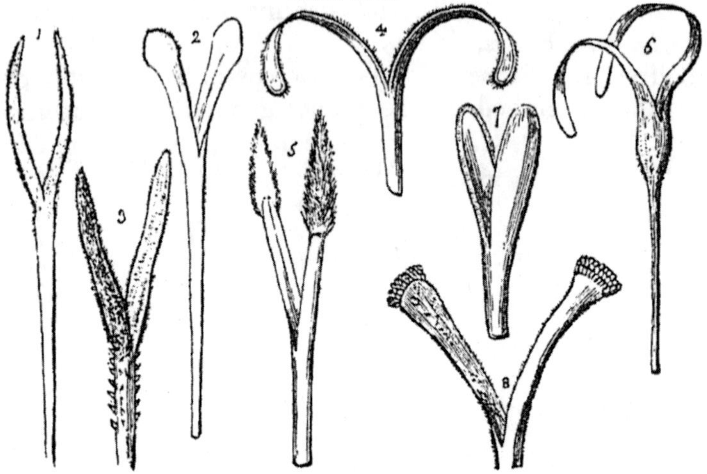

FIG. 143.

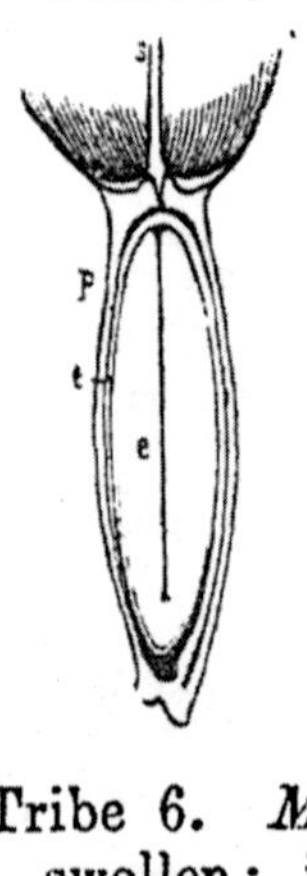

Fig. 142. Styles and stigmas of Composite Flowers to illustrate De Candolle's tribes, after Heyland and Lindley. 1. *Albertinia erythropappa* (Vernonieæ). 2. *Anisochæta mikanioides* (Eupatorieæ). 3. *Blumea senecioides* (Asteroideæ). 4. *Mendezia bicolor* (Senecioideæ). 5. *Lipochæta umbellata* (Senecioideæ). 6. *Aplotaxis nepalensis* (Cynareæ). 7. *Leucomeris spectabilis* (Mutisieæ). 8. *Leuceria tenuis* (Nassauvieæ).——*Fig.* 143. Vertical section of the ripe fruit of the Groundsel (*Senecio vulgaris*), surmounted by a portion of the style, *s*; and the pappose limb of the calyx. *p*. Pericarp. *t*. Testa. *e*. Seed.

Tribe 6. *Mutisieæ.*—Style cylindrical or somewhat swollen; its arms usually blunt or truncate, very convex on the outside, and either covered at the upper part by a fine uniform hairiness, or absolutely free from hairs (*fig.* 142, 7). *Illustrative Genera :*—Mutisia, *Linn. fil.*; Printzia, *Cass.*

Tribe 7. *Nassauvieæ.*—Style never swollen; its arms, long, linear, truncate, and fringed only at the point (*fig.* 142, 8). *Illustrative Genera*:—Nassauvia, *Juss.*; Trixis, *P. Br.*

SUB-ORDER 3. *Liguliforæ* or *Cichoraceæ.*—Florets all ligulate and perfect. Juice milky.

Tribe 8. *Cichoreæ.*—Style cylindrical at the upper part; its arms somewhat obtuse, and equally pubescent (*fig.* 133, *s*). *Illustrative Genera*:—Cichorium, *Linn.*; Taraxacum, *Haller.*

2. Hypostamineæ.

ERICACEÆ, the Heath Order.—Character.—*Shrubs* or small *trees*. *Leaves* entire, evergreen, opposite or whorled, exstipulate. *Calyx* 4—5-cleft, inferior, persistent (*fig.* 144, *c*). *Corolla* hypogynous, monopetalous (*figs.* 144, *t*, *l*, and 145), 4—5-cleft; *æstivation* imbricate. *Stamens* hypogynous (*figs.* 145 and 146), as many, or twice as many, as the divisions of the corolla; *anthers* 2-celled (*fig.* 148, *l*), opening by pores or slits (*fig.* 147, *r*), appendiculate (*fig.* 147, *a*). *Ovary* many-celled, with numerous ovules, surrounded by a disk or scales; *placentas* axile (*fig.* 145); *style* 1 (*figs.* 145 and 146); *stigma* simple or lobed. *Fruit* capsular or rarely baccate. *Seeds* numerous, small, anatropous; *embryo* in the axis of fleshy albumen.

FIG. 144.

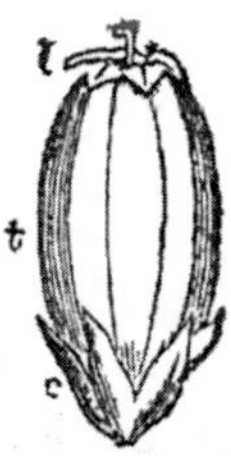

Fig. 144. Flower of a species of Heath (*Erica*). *c.* Calyx, within which is an urceolate corolla, *t*, *l*.

Diagnosis.—Shrubs or small trees. Leaves entire, evergreen, exstipulate. Calyx and corolla 4—5-merous. Calyx inferior. Corolla hypogynous, monopetalous. Stamens hypogynous; anthers 2-celled, appendiculate, with porous dehiscence. Ovary many-celled; style 1; placentas axile. Fruit with several cells. Seeds small, numerous, with fleshy albumen.

Division of the Order and Illustrative Genera.—The order has been divided into four tribes as follows:—

Tribe 1. *Arbuteæ.*—Corolla deciduous. Fruit baccate. *Illustrative Genus:*—Arbutus, *Linn.*

Tribe 2. *Andromedeæ.*—Buds usually clothed with scales. Corolla deciduous. Fruit capsular, loculicidal. *Illustrative Genus:* Andromeda, *Linn.*

FIG. 145.

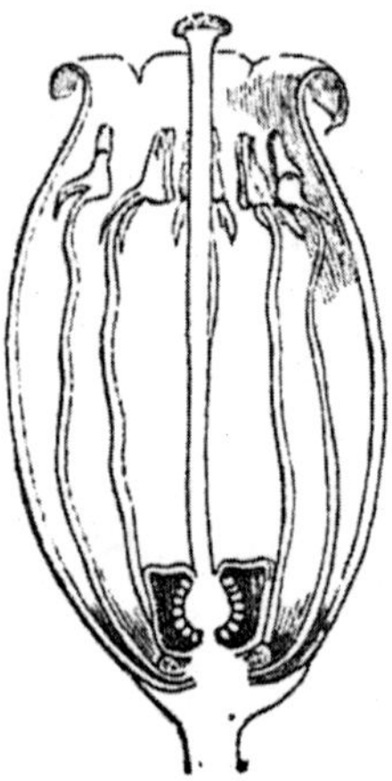

FIG. 146.

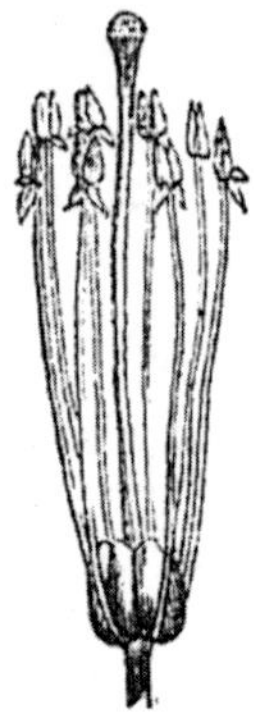

FIG. 147.

FIG. 148.

Fig. 145. Vertical section of the flower of a species of Heath (*Erica*).——*Fig.* 146. Essential organs of the same. The stamens are seen to be hypogynous.——*Fig.* 147. Appendiculate anther attached to filament, *f*, of the Fine-leaved Heath (*Erica cinerea*). *a.* Appendage. *l.* Lobes. *r.* Lateral slit where dehiscence takes place.——*Fig.* 148. Quadrifurcate 2-celled anther of *Gualtheria procumbens*, attached to filament. *f.* *l.* Anther lobes.

Tribe 3. *Ericeæ.*—Buds naked. Corolla persistent. Fruit capsular, usually loculicidal, or rarely septicidal. *Illustrative Genera:*—Erica, *Linn.*; Calluna, *Salisb.*

Tribe 4. *Rhodoreæ.*—Buds scaly, cone-like. Corolla deciduous. Fruit capsular, septicidal. *Illustrative Genera:*—Azalea, *Linn.*; Phyllodoce, *Salisb.*

3. Epipetalæ.

GENTIANACEÆ, the Gentian Order.—Character.—*Herbs* or rarely *shrubs*, usually smooth. *Leaves* gene rally simple, entire, opposite, sessile, and strongly ribbed; rarely alternate, or stalked, or compound; always exstipulate. *Flowers* (*fig.* 149) regular, solitary, and terminal, or in di–tri-chotomous cymes. *Calyx* inferior (*fig.* 149), persistent, usually with 5 divisions, or occasionally with 4, 6, 8, or 10. *Corolla* persistent, its divisions corresponding in number to those of the calyx; *æstivation* imbricate-twisted (*fig.* 149), or induplicate. *Stamens* as many as the segments of the corolla and alternate with them. *Ovary* 1-celled, or rarely partially 2-celled from the projection inwards

FIG. 149.

FIG. 150.

Fig. 149. Portion of the floral axis of a species of Gentian (*Gentiana acaulis*), terminated by a solitary flower, below which are two bracts.——*Fig.* 150. Fruit of a Gentian dehiscing in a septicidal manner.

of the placentas; *ovules* numerous; *placentas* 2, parietal (*fig.* 150), anterior and posterior to the axis, and frequently turned inwards; *style* 1; *stigmas* 2, right and left of the axis. *Fruit* capsular (*fig.* 150), 1—2-celled, 2-valved, with septicidal dehiscence; or rarely fleshy and indehiscent. *Seeds* numerous (*fig.* 150), small; *embryo* minute, in the axis of fleshy albumen.

Diagnosis.—Usually smooth herbs. Leaves without stipules. Flowers regular, solitary and terminal, or in cymes. Calyx and corolla persistent, with an equal number of lobes. Stamens alternate to the lobes of the corolla, and equal to them in number. Ovary superior, 1-celled, with 2 parietal placentas placed anterior and posterior, sometimes meeting in the centre and forming a 2-celled ovary; style 1; stigmas 2. Seeds small, numerous, with a minute embryo in the axis of fleshy albumen.

Division of the Order and Illustrative Genera:—The order may be divided into two sub-orders, the characters of which are essentially derived from the æstivation of the corolla:—

Sub-order 1. *Gentianeæ.*—Leaves opposite. Corolla imbricate-twisted (*fig.* 149). *Illustrative Genera*:—Gentiana, *Linn.*; Erythræa, *Pers.*; Chlora, *Linn.*

Sub-order 2. *Menyantheæ.*—Leaves alternate. Corolla induplicate. *Illustrative Genera*:—Menyanthes, *Tourn.*; Villarsia, *Vent.*

CONVOLVULACEÆ, the Convolvulus Order.—Cha-

FIG. 151.

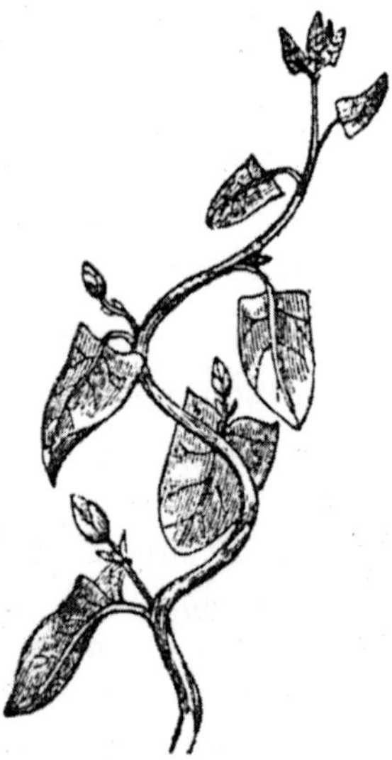

FIG. 152.

Fig. 151. Twining stem of a species of *Convolvulus.*——*Fig.* 152. Cuscuta or Dodder-plant.

racter.—*Herbs* or *shrubs*, generally twining (*fig.* 151), or trailing, and milky; sometimes leafless and parasitic (*fig.* 152). *Leaves* (*fig.* 151) alternate, exstipulate.

Calyx inferior (*fig.* 153), with 5 deep divisions (*fig.* 154), much imbricated, persistent. *Corolla* (*figs.* 153, 154 and 155) 5-partite or 5-plaited, regular, deciduous, sometimes with scales in its tube (*fig.* 155); *æstivation* plaited or imbricate. *Stamens* 5, alternate with the

FIG. 153.

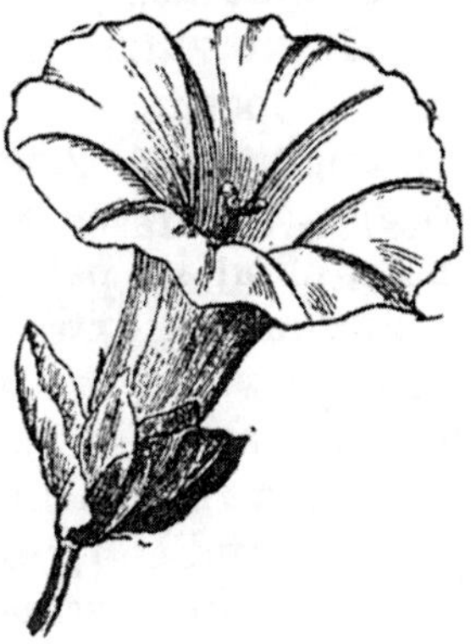

FIG. 154.

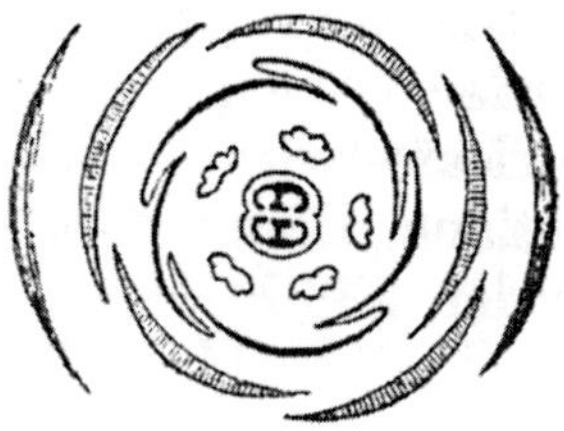

FIG. 155.

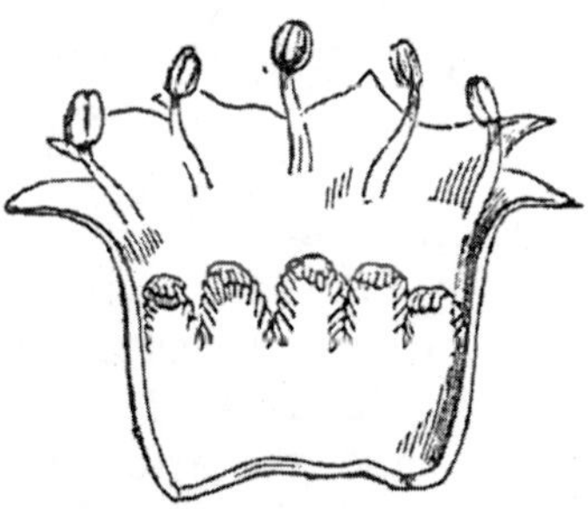

FIG. 156.

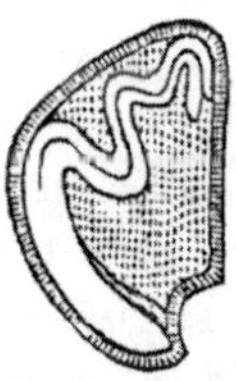

FIG. 157.

Fig. 153. Flower of the Great Bindweed (*Convolvulus sepium*).——*Fig.* 154. Diagram of the same flower, showing two bracts on the outside of the calyx.——*Fig.* 155. Corolla of a Dodder (*Cuscuta*) laid open to show the scales in its tube.——*Fig.* 156. Vertical section of the seed of *Convolvulus sepium.*——*Fig.* 157. Spiral embryo of a species of *Cuscuta.*

lobes of the corolla (*figs.* 154 and 155). *Ovary* (*fig.* 154) 2, 3, or 4-celled, or the carpels are more or less distinct; *ovules* 1—2 in each cell or carpel, erect. *Fruit* capsular, 1—4-celled, with septifragal dehiscence. *Embryo* large, curved or coiled in mucilaginous or fleshy albumen, with foliaceous plaited cotyledons (*fig.* 156); or in *Cuscuta* the embryo (*fig.* 157) is filiform, spiral, and the cotyledons scarcely perceptible; radicle inferior.

Diagnosis.—Generally twining or trailing milky herbs, with alternate exstipulate leaves; or parasitic and leafless. Calyx of 5 imbricate sepals, inferior. Corolla regular, 5-plaited or 5-lobed. Stamens 5, alternate with the lobes of the corolla. Ovary 2—4-celled. Fruit 2—4-celled, capsular, septifragal. Embryo curved, coiled, or spiral, in albumen; radicle inferior. *Illustrative Genera*:—Convolvulus, *Linn.*; Cuscuta, *Tourn.*

SOLANACEÆ, the Solanum or Nightshade Order.—Character.—*Herbs*, or rarely *shrubs* or *trees*, with a colourless juice. *Leaves* alternate, often in pairs, exstipulate. *Inflorescence* axillary, or frequently extra-axillary (*fig.* 158). *Flowers* isomerous (*figs.* 159 and 163). *Calyx* (*figs.* 159 and 163) with 5 or rarely 4 divisions, usually persistent, often growing during the ripening of the fruit (*accrescent*). *Corolla* (*figs.* 159 and 163) regular or somewhat irregular, 5- or rarely 4-partite; *æstivation* valvate, induplicate, plicate, or imbricate. *Stamens* equal in number to the lobes of the corolla, with which they are alternate (*figs.* 159 and 163); *anthers* 2-celled, sometimes connate above (*fig.* 160, *e*), dehiscing longitudinally (*fig.* 162), or by terminal pores (*figs.* 160 and 161). *Ovary* superior (*figs.* 160, *o*, and 162), usually 2-celled (*fig.* 163), in which case the cells are placed anterior and posterior, rarely 3- to 5-celled; *placentas* axile (*figs.* 160, *o*, and 163); *style* (*fig.* 164, *s*) simple; *stigma* simple (*fig.* 164, *g*) or 2-lobed. *Fruit* capsular (*fig.* 165) or baccate, 2- or more celled. *Seeds* numerous, albuminous (*fig.* 166, *alb*); *embryo* straight, or usually curved in a more or less annular or spiral form (*fig.* 166).

Diagnosis.—Herbs, or rarely shrubs or trees, with alternate exstipulate leaves, and a colourless juice. Flowers isomerous. Calyx and corolla with 5, or rarely 4 divisions. Corolla regular or very slightly irregular; æstivation valvate, imbricate, induplicate, or plaited. Stamens equal in number to the lobes of the corolla, with which they are alternate; anthers 2-celled, with porous or longitudinal dehiscence. Ovary superior, with axile

FIG. 158.

FIG. 159.

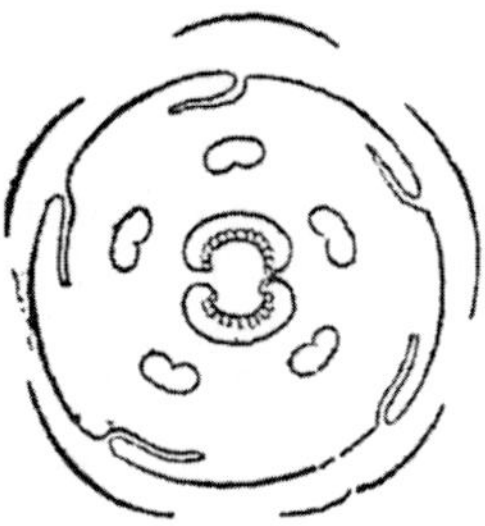

FIG. 160.

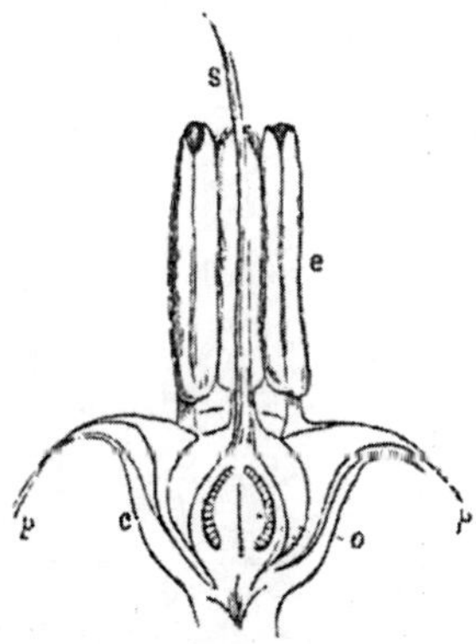

FIG. 161.

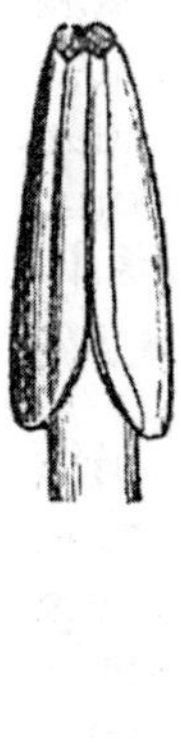

FIG. 162.

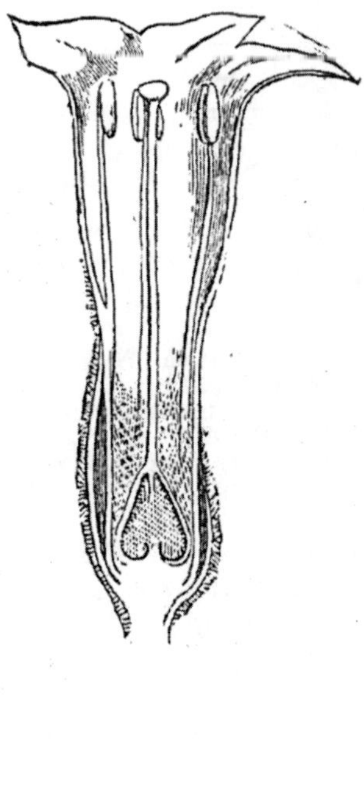

Fig. 158. A portion of the stem of the Woody Nightshade (*Solanum Dulcamara*), bearing flowering stalk and an auriculate leaf.——*Fig.* 159. Diagram of the flower of the Potato (*Solanum tuberosum*)——*Fig.* 160. Vertical section of the flower of the same. *c.* Calyx. *p, p.* Corolla. *o.* Ovary. *e.* Stamens. *s.* Style and stigma.——*Fig.* 161. Anther lobes of a species of *Solanum* opening by pores at the apex.——*Fig.* 162. Vertical section of the flower of Tobacco (*Nicotiana Tabacum*).

placentation, usually 2-celled, the cells being then placed anterior and posterior; or rarely more celled. Fruit

FIG. 163.

FIG. 164.

FIG. 165.

FIG. 166.

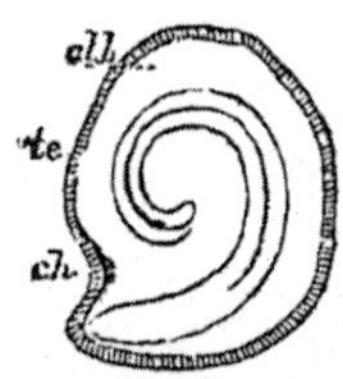

Fig. 163. Diagram of the flower of Tobacco (*Nicotiana Tabacum*). ——*Fig.* 164. Compound syncarpous pistil of the same. *t.* Thalamus. *o.* Ovary. *s.* Style. *g.* Capitate stigma.——*Fig.* 165. Capsule of *Datura Stramonium*, showing septifragal dehiscence. ——*Fig.* 166. Vertical section of the seed of *Solanum Dulcamara*. *te.* Testa. *ch.* Chalaza. *alb.* Albumen.

dehiscent or indehiscent, 2- or more celled. Seeds numerous, albuminous. *Illustrative Genera* :—Solanum, *Tourn.*; Hyoscyamus, *Tourn.*; Atropa, *Linn.*

FIG. 167.

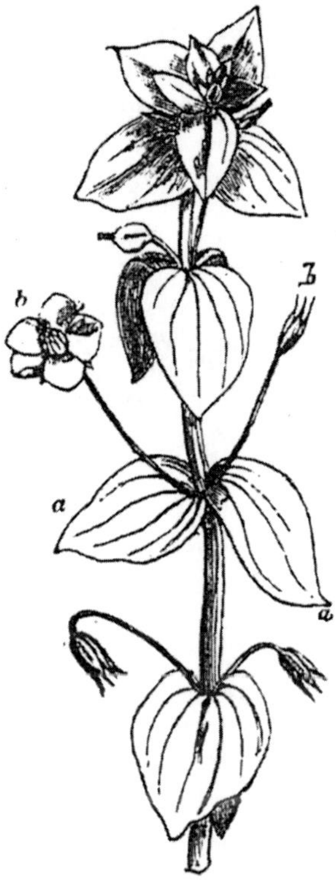

FIG. 168.

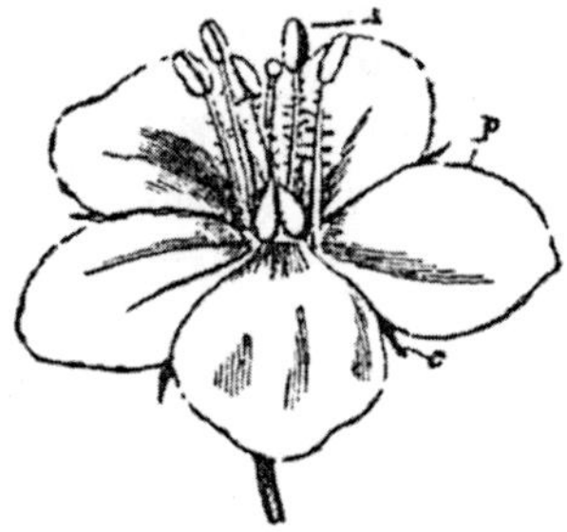

FIG. 169.

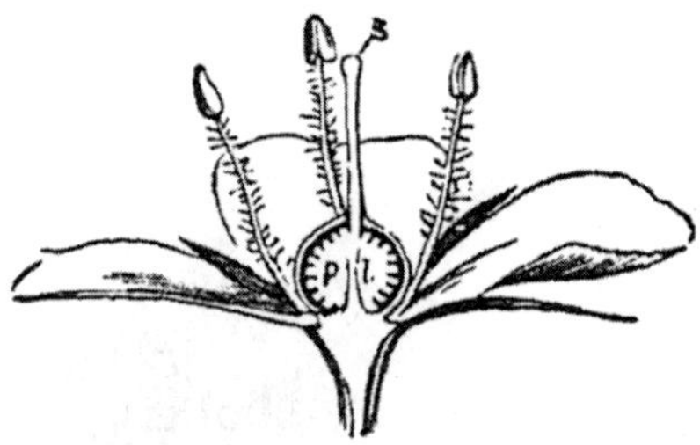

FIG. 170.

FIG. 171.

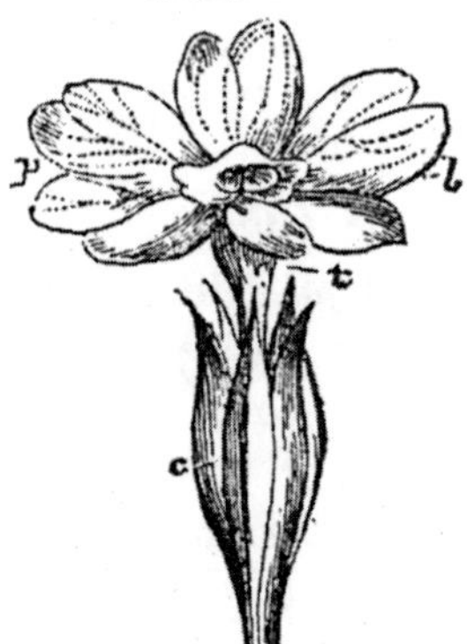

Fig. 167. Flowering stalk of the Pimpernel (*Anagallis arvensis*). *b, b.* Solitary flowers arising from the axil of leafy bracts, *a, a.*——*Fig.* 168. Flower of the same. *c.* Calyx. *p.* Petals. *s.* Stamens.——*Fig.* 169. Vertical section of the flower. *pl.* Free central placenta. *s.* Style and capitate stigma.——*Fig.* 170. Partite inferior calyx of the flower of the same.——*Fig.* 171. Flower of a species of *Primula*. *c.* Calyx, within which is seen a hypocrateriform corolla, *p.* *t.* Tube of the corolla. *l.* Limb.

PRIMULACEÆ, the Primrose Order.—Character.—*Herbs.* *Leaves* (*fig.* 167, *a*, *a*) cauline, and then simple, opposite, whorled, or rarely alternate, and exstipulate; or radical. *Flowers* regular, perfect (*fig.* 168). *Calyx* (*figs.* 170 and 171, *c*), generally 5-

FIG. 172.

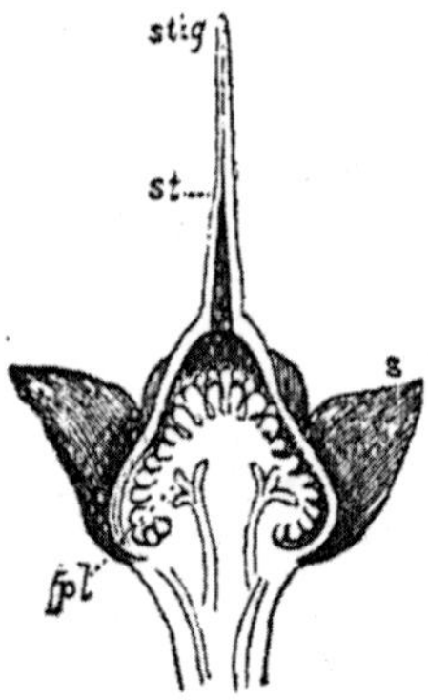

FIG. 173.

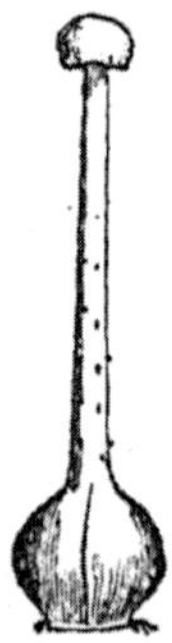

FIG. 174.

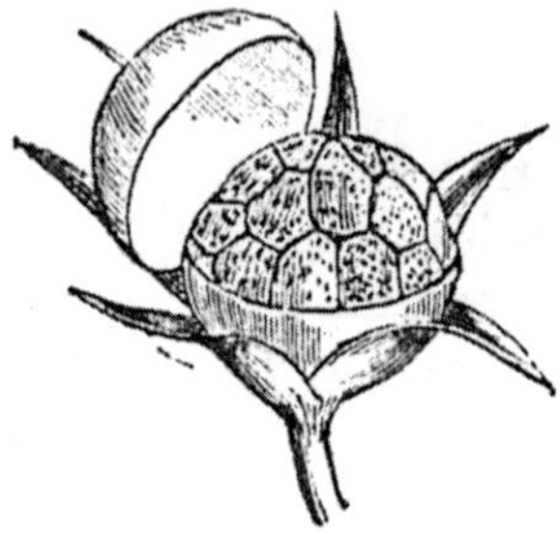

FIG. 175.

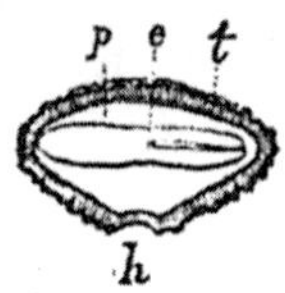

Fig. 172. Vertical section of the pistil of *Cyclamen europæum.* *s.* Sepals. *pl.* Free central placenta. *st.* Style. *stig.* Stigma.——*Fig.* 173. Pistil of the Primrose (*Primula vulgaris*), composed of several united carpels, and hence termed compound and syncarpous. There is but one style surmounted by a capitate stigma.——*Fig.* 174. Pyxis or fruit of the Pimpernel (*Anagallis arvensis*).——*Fig.* 175. Vertical section of the seed of *Primula elatior.* *t.* Integuments. *p.* Albumen. *e.* Embryo. *h.* Hilum.

or rarely 4—9-cleft, persistent, inferior (*fig.* 170), or semi-superior in *Samolus.* *Corolla* (*figs.* 168, *p*, and 171, *l*), usually 5- or rarely 4—9-cleft, very rarely absent, or rarely of distinct petals. *Stamens* (*fig.* 168, *s*) equal in number to the segments of the corolla or separate petals, and opposite to them, or in apetalous flowers hypogynous and alternating with the divisions of the calyx. *Ovary* superior (*figs.* 169 and 170), or rarely partly inferior, 1-celled (*figs.* 169 and 172); *placenta* free central (*figs.* 169, *pl*, and 172, *pl*); *style* 1 (*figs.*

170 and 173); *stigma* capitate (*figs.* 169, *s*, 170, and 173). *Fruit* a capsule, dehiscing transversely and forming a pyxis (*fig.* 174), or in a valvular manner. *Seeds* numerous, with fleshy or horny albumen (*fig.* 175, *p*); *embryo* placed transversely to the hilum (*fig.* 175, *e*).

Diagnosis.—Herbs with simple, exstipulate, cauline or radical leaves, and regular perfect flowers. Stamens equal in number to the lobes of the corolla or separate petals, and opposite to them. Ovary superior, 1-celled, with a free central placenta; style 1; stigma capitate. Fruit capsular, with transverse or longitudinal dehiscence. Seeds numerous, with albumen, and the embryo parallel to the hilum. *Illustrative Genera*:—Primula, *Linn.*; Anagallis, *Tourn.*; Glaux, *Tourn.*; Samolus, *Tourn.*

BORAGINACEÆ, the Borage Order.—Character.—*Herbs* or *shrubs*, with more or less rounded, usually

FIG. 176.

Fig. 176. Scorpioid cyme of Comfrey (*Symphytum officinale*).

rough, and hairy stems. *Leaves* alternate, entire or rarely sinuate, usually rough, exstipulate (*fig.* 176).

Inflorescence scorpioid (*figs.* 176 and 177). *Flowers* regular, symmetrical (*fig.* 178). *Calyx* persistent, inferior (*fig.* 179), 5-partite or lobed (*fig.* 178). *Corolla* (*figs.* 178 and 180) regular or nearly so, 5-partite or lobed, usually with scales in its throat (*fig.* 180, *r*); *æstivation* imbricate (*fig.* 178). *Stamens* (*fig.* 178) equal in number to the lobes of the corolla and alternate with them. *Ovary* superior and composed of 2 carpels, each of which is 2-lobed and 2-celled (*fig.* 181, *ov*), with a

FIG. 177.

FIG. 178.

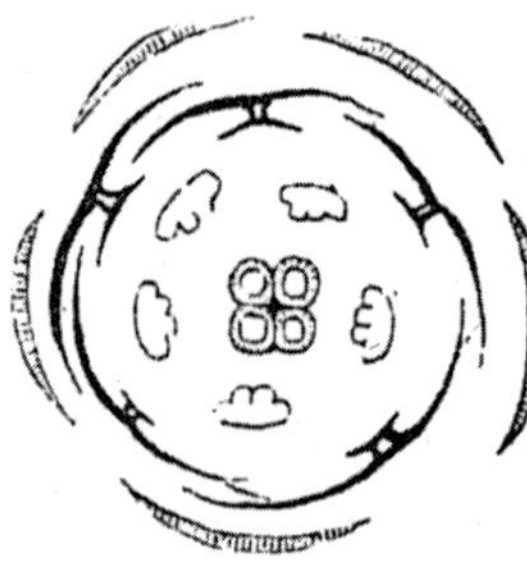

FIG. 179.

Fig. 177. Scorpioid cyme of the Forget-me-not (*Myosotis palustris*).——*Fig.* 178. Diagram of the flower of the Comfrey (*Symphytum officinale*).——*Fig.* 179. Vertical section of the calyx and fruit of a species of *Myosotis*. Two achænia are seen, and two have been removed.

solitary ovule in each cell; *style* 1 (*fig.* 181, *d*), basilar; *stigma* simple or bifid. *Fruit* consisting of usually 4 distinct achænia, placed at the bottom of the persistent calyx (*figs.* 179 and 182). *Seeds* exalbuminous; *embryo* straight, with a superior radicle.

Diagnosis.—Herbs with rounded usually rough stems, and alternate exstipulate leaves. Inflorescence scorpioid. Flowers regular and perfect. Calyx, corolla, and stamens equal in number, the latter being alternate with the divisions of the corolla. Ovary superior, deeply

4-lobed, with one ovule in each lobe; style 1, basilar. Fruit composed of 4 achænia placed at the bottom of the persistent calyx. Seeds without albumen. *Illustrative Genera:*—Echium, *Tourn.*; Symphytum, *Tourn.*; Myosotis, *Linn.*

FIG. 180.

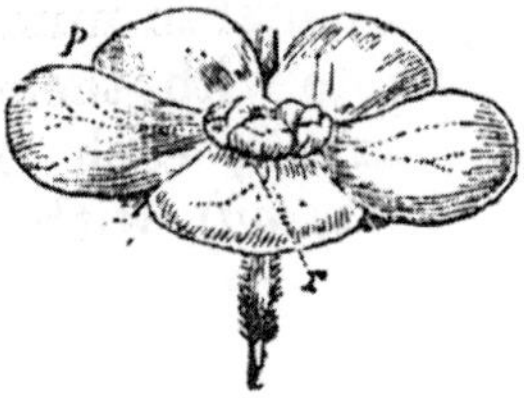

FIG. 181.

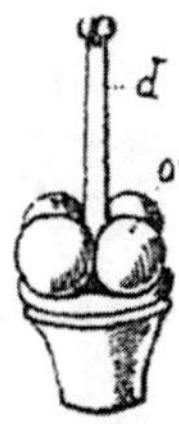

FIG. 182.

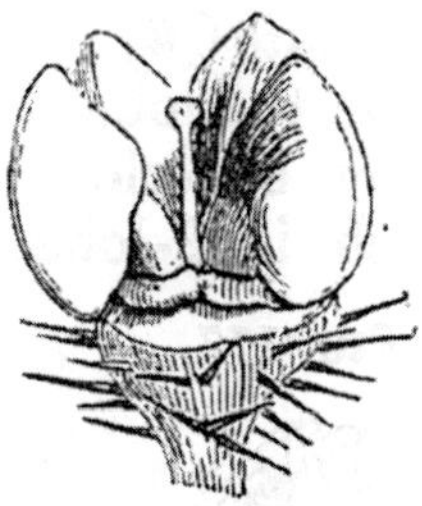

Fig. 180. Flower of the Forget-me-not (*Myosotis palustris*). *p.* Rotate corolla. *r.* Scales projecting from its limb close to the throat.——*Fig.* 181. Pistil of the same. *ov.* 4-lobed ovaries. *d.* Styles united.——*Fig.* 182. Achænia of Bugloss (*Lycopsis arvensis*).

LABIATÆ, the Labiate Order.—Character.—*Herbs* (*fig.* 183) or *shrubby plants*, with usually square stems. *Leaves* opposite (*fig.* 183) or whorled, commonly strong-scented, entire or divided, exstipulate. *Flowers* generally in axillary cymes, which are arranged in a somewhat whorled manner so as to form what are called verticillasters (*fig.* 183). *Calyx* inferior, persistent, either tubular (*fig.* 185), 5- or 10-toothed, and regular, or nearly so, or irregular and bilabiate (*fig.* 186), with 3—10 divisions; the odd tooth or division always posterior (*figs.* 184 and 185). *Corolla* (*figs.* 186–190) usually more or less bilabiate, with the upper lip undivided (*figs.* 187 and 190), or bifid (*figs.* 186 and 188), and commonly more or less arched over the lower lip (*fig.* 187), or sometimes nearly suppressed (*fig.*189); the lower lip 3-lobed (*fig.* 189), with the odd lobe anterior (*figs.* 184 and 189); or rarely the corolla is nearly regular. *Stamens* usually 4, and then commonly didynamous (*figs.* 190 and 191), or very rarely of nearly equal length, or only 2 by abortion (*figs.* 186 and

FIG. 183.

Fig. 183. Flowering stalk of the White Dead-nettle (*Lamium album*).

FIG. 184.

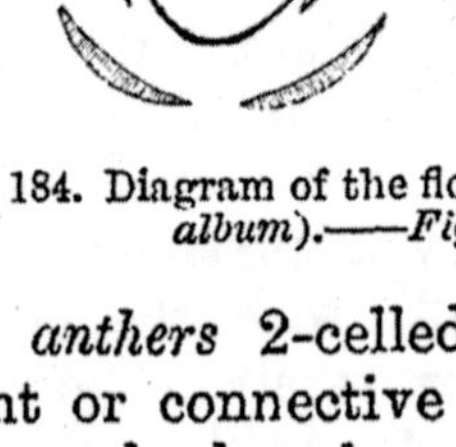

FIG. 185

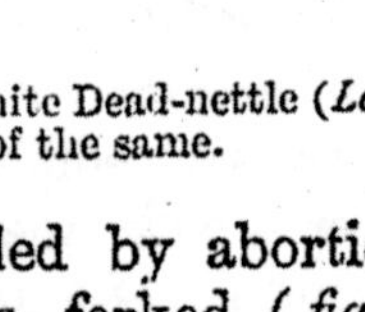

Fig. 184. Diagram of the flower of the White Dead-nettle (*Lamium album*).—*Fig.* 185. Calyx of the same.

192); *anthers* 2-celled or 1-celled by abortion; the filament or connective sometimes forked (*fig.* 192), each branch then bearing a perfect cell, or the cell on one side obsolete or sterile (*fig.* 193, *ls*). *Ovary* deeply 4-lobed (*fig.* 194), imbedded in the thalamus, and formed of two carpels, each of which has 2 deep lobes, and each lobe with one erect ovule; *style* 1, basilar

FIG. 186.

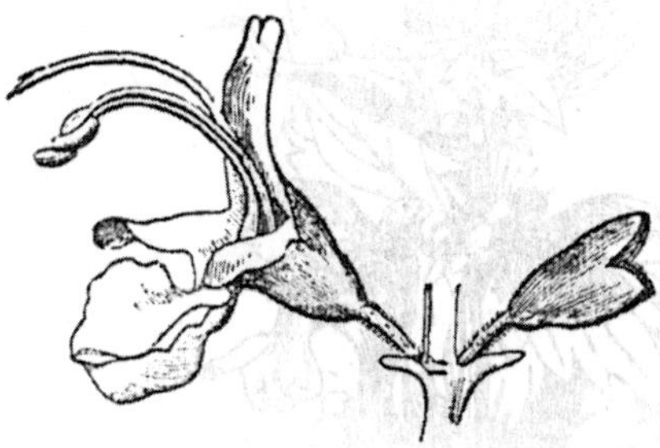

FIG. 187.

Fig. 186. Flower of Rosemary (*Rosmarinus officinalis*) with upper lip of corolla bifid; the right-hand flower showing bilabiate calyx.——*Fig.* 187. Ringent or gaping corolla of Dead-nettle (*Lamium album*), showing the entire upper lip and trifid lower lip.

(*fig.* 194); *stigma* bifid (*fig.* 194). *Fruit* composed of from 1—4 achænia, enclosed by the persistent calyx. *Seed* erect, with little or no albumen; *embryo* erect, with flat *cotyledons*; *radicle* inferior.

FIG. 188.

FIG. 189.

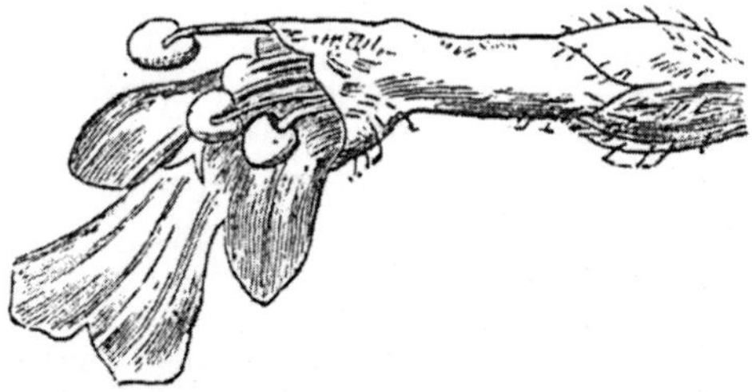

Fig. 188. Back view of the flower of a species of *Teucrium*, showing the bifid upper lip of the corolla.——*Fig.* 189. Flower of the common Bugle (*Ajuga reptans*).

Diagnosis.—Herbs or shrubby plants, with opposite exstipulate leaves and commonly square stems. Flowers irregular, unsymmetrical. Calyx persistent.

Corolla usually more or less bilabiate. Stamens generally four and then commonly didynamous, or rarely of

FIG. 190.

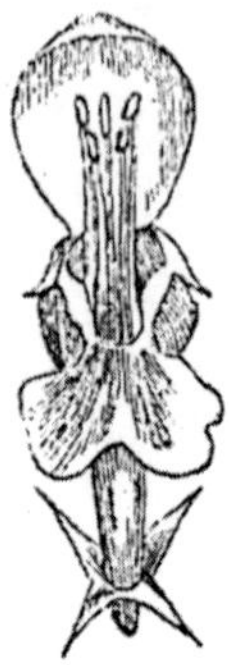

FIG. 191.

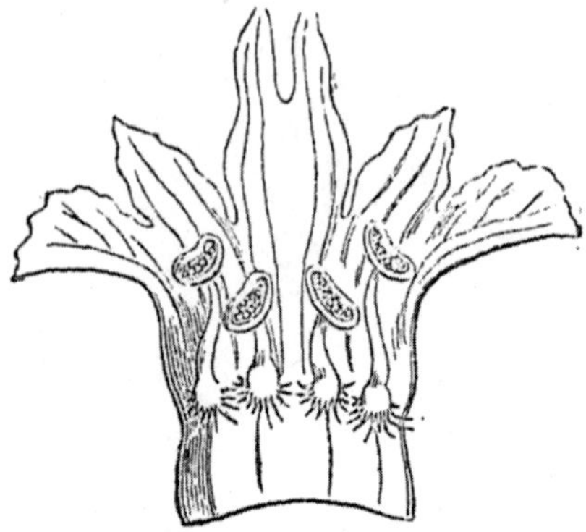

Fig. 190. Front view of the flower of a species of *Lamium.*——*Fig.* 191. The corolla of the Horehound (*Marrubium vulgare*) cut open.

FIG. 192.

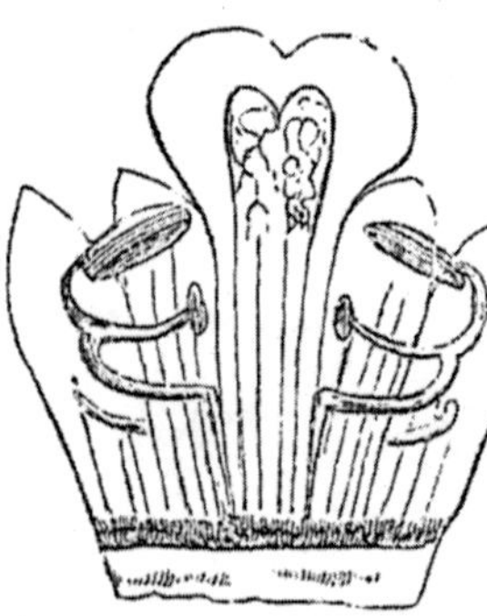

FIG. 193.

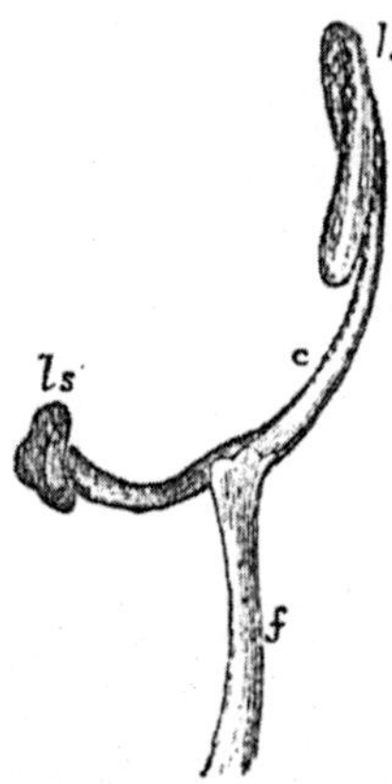

FIG. 194.

Fig. 192. The corolla of the Garden Sage (*Salvia officinalis*) cut open.——*Fig.* 193. Stamen of the same. *f.* Filament. *c.* Connective, bearing at one end a cell, *lf*, containing pollen, which is then said to be fertile; and at the other end a sterile cell, *ls*, or one without pollen.——*Fig.* 194. Lobed ovary, style, and bifid stigma of the same.

equal length; or only 2 by abortion. Ovary deeply 4-lobed; style 1, basilar; stigma bifid. Fruit consisting of from 1—4 achænia, enclosed by the persistent calyx.

Seed erect, with little or no albumen. *Illustrative Genera*:—Mentha, *Linn.*; Salvia, *Linn.*; Lamium, *Linn.*; Marrubium, *Linn.*

SCROPHULARIACEÆ, the Figwort Order.—Character.—*Herbs* or rarely *shrubby plants*, with alternate, opposite, or whorled *leaves*, generally exstipulate, or rarely with stipules; sometimes parasitical on roots. *Inflorescence* axillary. *Flowers* (*figs.* 195 and 196) anisomerous, irregular. *Calyx* inferior (*fig.* 199), persistent (*fig.* 201), 4—5-partite (*figs.* 195 and 199). *Corolla* more (*figs.* 197 and 198), or less (*figs.* 196 and

FIG. 195.

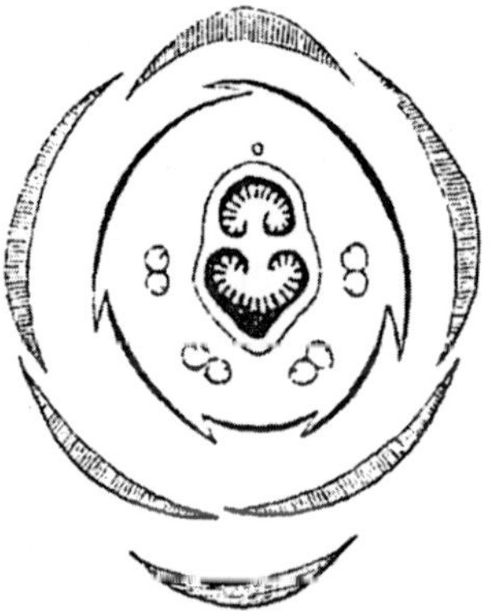

FIG. 196.

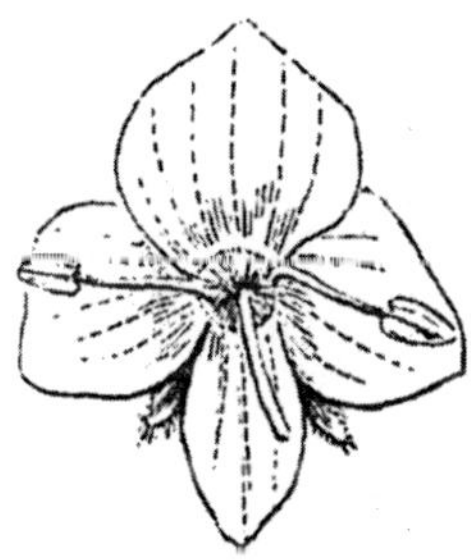

Fig. 195. Diagram of the flower of the Great Snapdragon (*Antirrhinum majus*), with one bract below.——*Fig.* 196. Flower of a species of Speedwell (*Veronica*).

199) irregular, sometimes gibbous (*fig.* 197) or spurred (*fig.* 198), 4—5-partite (*fig.* 195); *æstivation* imbricate (*fig.* 195). *Stamens* 2 (*fig.* 196) or 4, in the latter case didynamous (*fig.* 200), rarely 5 or with a rudimentary fifth; *anthers* 1- or 2-celled. *Ovary* superior, usually 2-celled with axile placentation (*fig.* 195), its component carpels being placed anterior and posterior; *style* 1 (*fig.* 196); *stigma* undivided (*fig.* 196) or 2-lobed. *Fruit* usually capsular (*fig.* 201), with variable dehiscence, or rarely baccate, usually 2-celled; *placentas* axile. *Seeds* generally numerous, small, albuminous, *embryo* straight or slightly curved.

Diagnosis.—Herbs or rarely shrubs. Flowers irregular, anisomerous. Inflorescence axillary. Calyx and corolla with 4 or 5 divisions. Corolla more or less

FIG. 197.

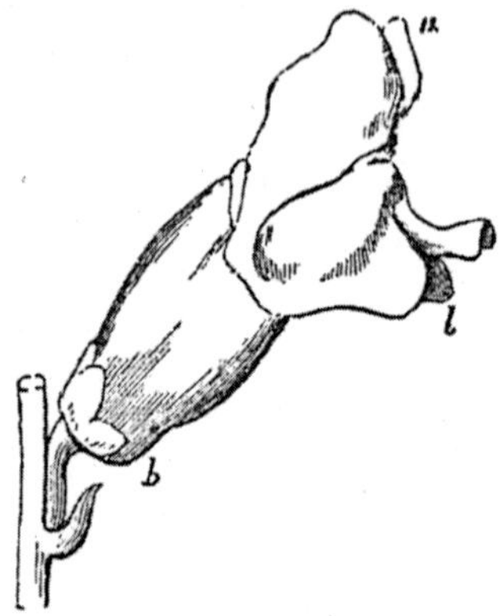

FIG. 198.

Fig. 197. Personate corolla of the Great Snapdragon (*Antirrhinum majus*). *l.* Lower lip. *u.* Upper lip. *b.* Gibbous base.——*Fig.* 198. Personate corolla of the common Toadflax (*Linaria vulgaris*), spurred at its base.

FIG. 199.

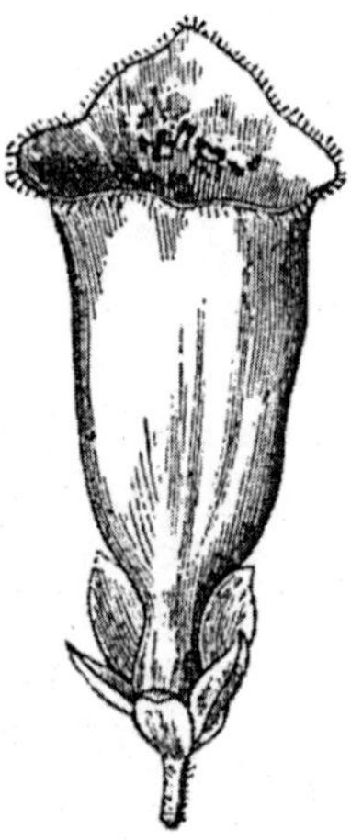

FIG. 200.

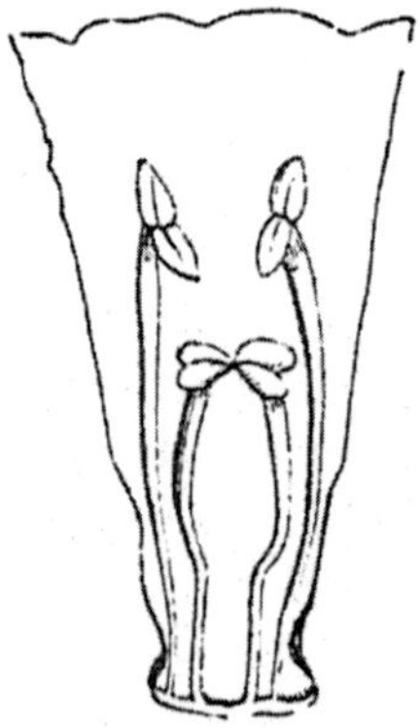

FIG. 201.

Fig. 199. Calyx and corolla of the Foxglove (*Digitalis purpurea*). ——*Fig.* 200. Didynamous stamens of the same.——*Fig.* 201. Fruit or capsule of a species of *Scrophularia*, dehiscing in a septicidal manner, and surrounded by the persistent calyx.

irregular, æstivation imbricate. Stamens 2 or 4 and then didynamous, or rarely 5; anthers 1- or 2-celled. Ovary superior, usually 2-celled, the cells placed anterior and posterior, with axile placentation; style 1. Fruit capsular or rarely baccate. Seeds generally numerous, albuminous. *Illustrative Genera*:—Verbascum, *Linn.*; Antirrhinum, *Tourn.*; Linaria, *Tourn.*; Scrophularia, *Linn.*; Veronica, *Tourn.*; Digitalis, *Linn.*

Sub-class IV. *Monochlamydeæ* or *Incompletæ*.

POLYGONACEÆ, the Buckwheat Order.—Character. *Herbs* or rarely *shrubs*. *Leaves* alternate, simple, com-

FIG. 202.

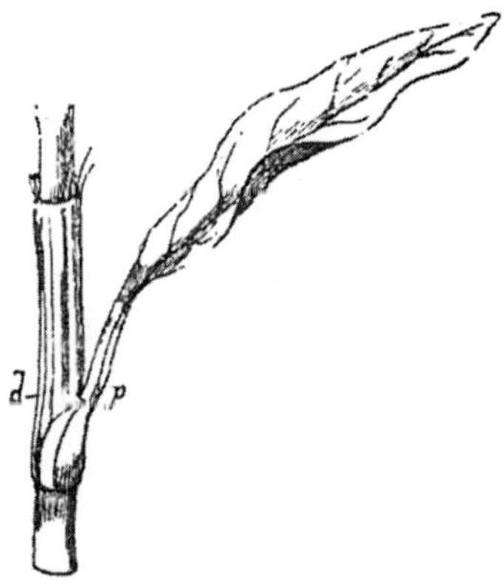

FIG. 203.

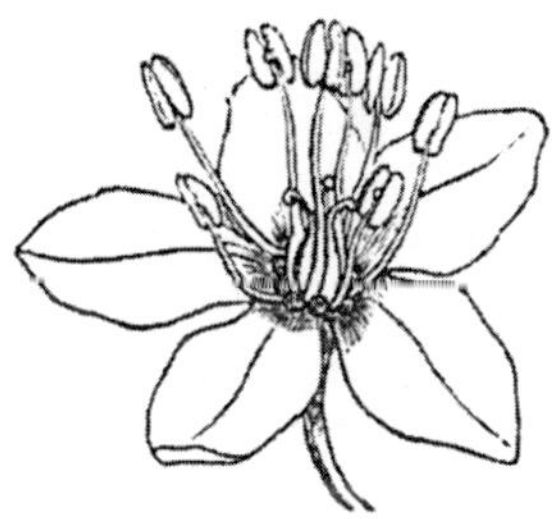

FIG. 204.

FIG. 205.

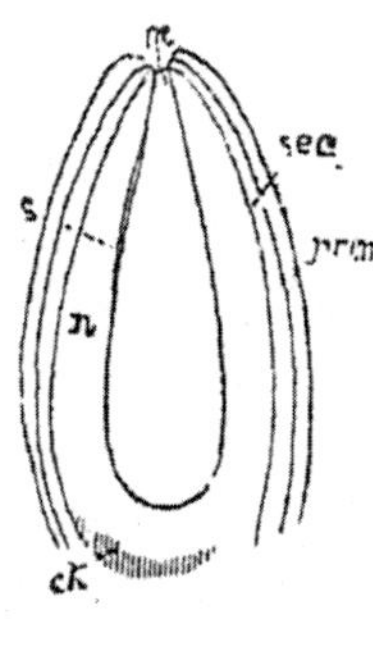

Fig. 202. Leaf and piece of stem of *Polygonum Hydropiper*. *l.* Lamina or blade. *p.* Petiole. *d.* Ochreaceous or sheathing stipules.——*Fig.* 203. Flower of a species of *Polygonum*.——*Fig.* 204. Pistil of a species of *Rumex*.——*Fig.* 205. Vertical section of the orthotropous ovule of a species of *Polygonum*. *ch.* Chalaza. *prim.* Primine. *sec.* Secundine. *n.* Nucleus. *s.* Embryo sac. *m.* Micropyle.

monly with ochreate stipules (*fig.* 202) above the swollen joints of the stem, or rarely exstipulate. *Flowers* perfect (*fig.* 203) or sometimes unisexual. *Calyx*[1] inferior (*fig.* 203), of from 3—6 sepals, more or less persistent, imbricate. *Stamens* few (*fig.* 203), hypogynous or rarely perigynous; *anthers* dehiscing longitudinally. *Ovary* superior (*fig.* 203), 1-celled; *styles* and *stigmas* 2—3 (*figs.* 203 and 204); *ovule* solitary, erect, orthotropous (*fig.* 205). *Fruit* usually a triangular nut, and commonly enveloped in the persistent calyx (*fig.* 206, *p*). *Seed* solitary, erect (*figs.* 206 and 207), and generally

FIG. 206.

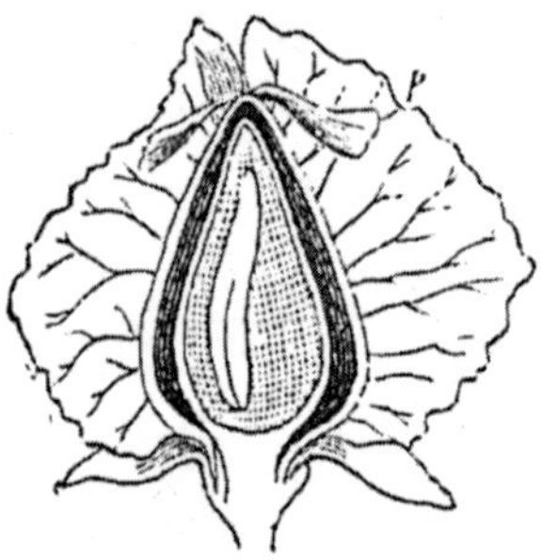

FIG. 207.

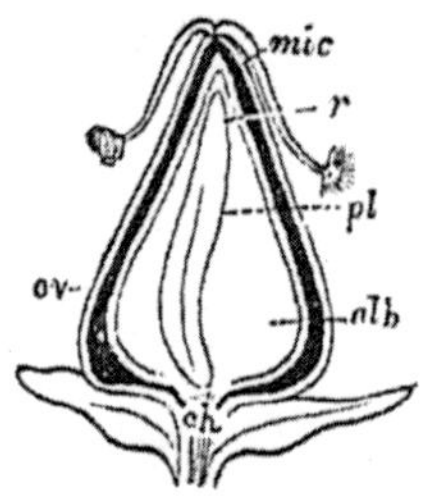

Fig. 206. Vertical section of the fruit of a species of *Rumex*. *p*. Enlarged calyx surrounding the fruit. The fruit contains a single erect orthotropous seed. The embryo is inverted or antitropous.——*Fig.* 207. Vertical section of the fruit and solitary erect orthotropous seed of a species of Dock (*Rumex*), with the enlarged calyx removed. *ov.* Pericarp. *mic.* Micropyle. *pl.* Embryo, which is inverted or antitropous, and turned towards one side of the albumen, *alb*. *ch.* Chalaza. *r.* Radicle.

with mealy or farinaceous *albumen*; *embryo* (*figs.* 206 and 207, *pl*), antitropous, with a superior radicle.

Diagnosis.—Usually herbs with ochreate stipules. Leaves simple, alternate. Calyx inferior, persistent, imbricate. Stamens definite. Ovary 1-celled; styles and stigmas 2—3. Fruit triangular. Seed solitary, erect, usually with mealy albumen; radicle superior.

[1] When there is but one floral envelope in Dicotyledonous plants, we call that the *calyx*, whatever be its colour or other peculiarity, in which nomenclature we follow the example of Lindley. By many botanists, however, the term *perianth* is employed in such cases, but we use that name only in speaking of Monocotyledonous plants.

Illustrative Genera :—Polygonum, *Linn.*; Rumex, *Linn.*

EUPHORBIACEÆ, the Spurge Order.—Character.—*Trees*, *shrubs*, or *herbs*, usually with an acrid milky juice. *Leaves* alternate or opposite, simple (*fig.* 208) or rarely compound, and with or without stipules. *Flowers* unisexual (*figs.* 209 and 210), monœcious (*fig.* 211) or diœcious, axillary or terminal, sometimes enclosed in a calyx-like involucre (*fig.* 211, *i*); achla-

FIG. 208.

FIG. 209.

FIG. 210.

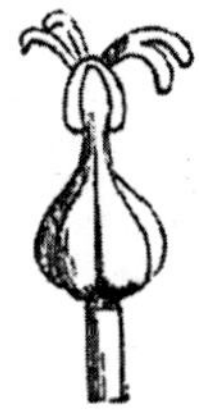

Fig. 208. Palmatifid leaf of the Castor-oil Plant (*Ricinus communis*).——*Fig.* 209. A male naked or achlamydeous flower of a species of *Euphorbia*, consisting of a solitary stamen, with a bract at the base of the flower-stalk.——*Fig.* 210. Naked or achlamydeous pistillate flower of a species of *Euphorbia*, with three forked styles.

mydeous (*figs.* 209 and 210), or with a lobed (*figs.* 212, *c*, and 213, *c*) inferior calyx, having on its inside glandular or scaly appendages (*fig.* 213, *t*, and 211, *b*), or even evident petals (*figs.* 212, *p*, and 213, *p*), which are either distinct or united. *Male flowers* consisting of 1 (*figs.* 209 and 211, *fm*) or more *stamens* (*figs.* 212, *e*, and 214), distinct or united into one (*fig.* 212, *a*) or more bundles (*fig.* 214); *anthers* 2-celled (*fig.* 209). *Female flowers* with a superior *ovary* (*figs.* 213, *o*, and 215), which is either elevated upon a stalk (*fig.* 211, *ff*) or sessile (*fig.* 215), 1-2-3- or many-celled; *styles* either absent or corresponding in number to the cells of the

FIG. 211.

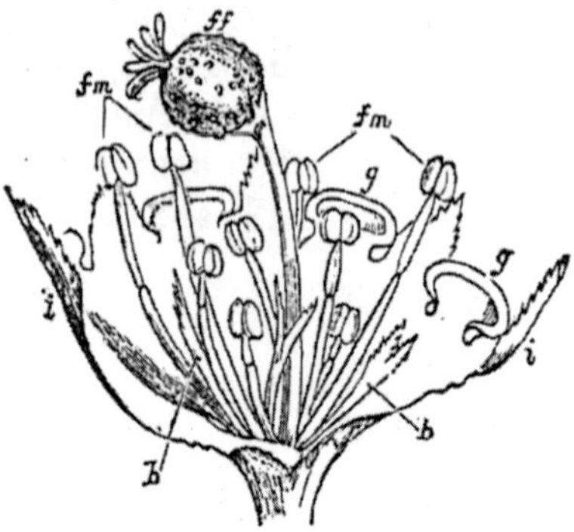

FIG. 212.

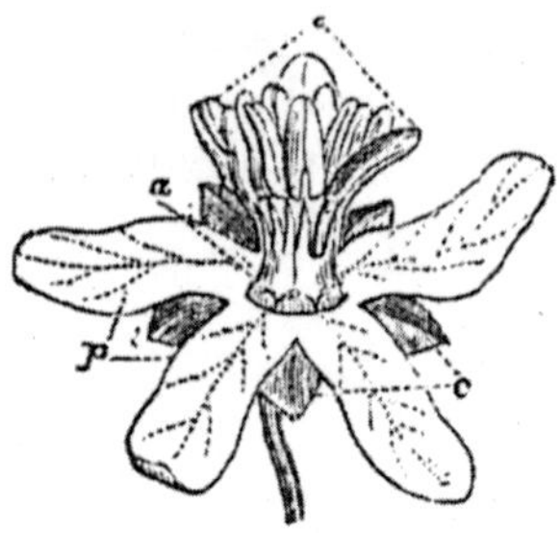

Fig. 211. Monœcious head of flowers of a species of *Euphorbia*. *i*. Involucre, a portion of which has been removed in front. *g*, *g*. Glands on the divisions of the involucre. *b*, *b*. Scales or bracts at the base of the flowers. *fm*, *fm*. Male flowers, each consisting of a stamen supported on a pedicel, to which it is articulated. *ff*. Female flower, supported on a stalk. From Jussieu.——*Fig.* 212. Male flower of *Jatropha Curcas*. *c*. Calyx. *p*. Corolla. *e*. Stamens united by their filaments into a tube, *a*, which occupies the centre of the flower, as there is no pistil.

FIG. 213.

FIG. 214.

Fig. 213. Female flower of one of the *Euphorbiaceæ*. *c*. Calyx. *p*, *p*. Petals. *t*. Membranous expansion round the ovary. *o*. Ovary with three styles, *s*, each of which is twice forked.——*Fig.* 214. One of the branched bundles of stamens of the Castor-oil Plant (*Ricinus communis*). *f*. United filaments.

ovary, entire or divided (*figs.* 210, 213, *s*, and 215); *stigmas* equal in number to the cells of the ovary, or, when the styles are divided, corresponding in number to their divisions (*figs.* 210, 213 and 215); *ovules* 1 or 2 in each cell, suspended from the inner angles. *Fruit* either dry, and its component carpels then separating from each other and from the axis (*figs.* 216, *c*, and 217) and usually opening with elasticity; or succulent and indehiscent. *Seeds* 1 or 2 in each cell, suspended

FIG. 215.

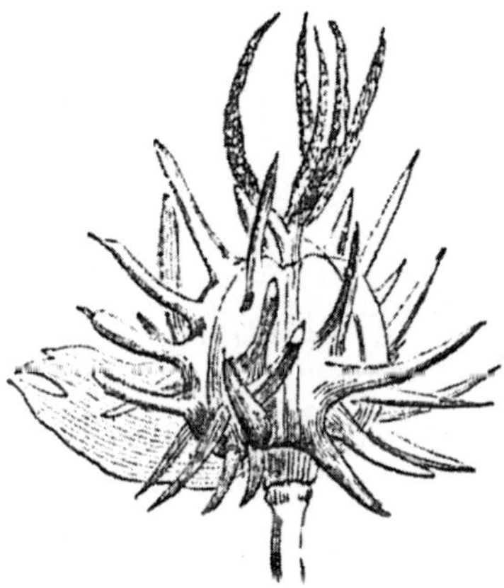

FIG. 216.

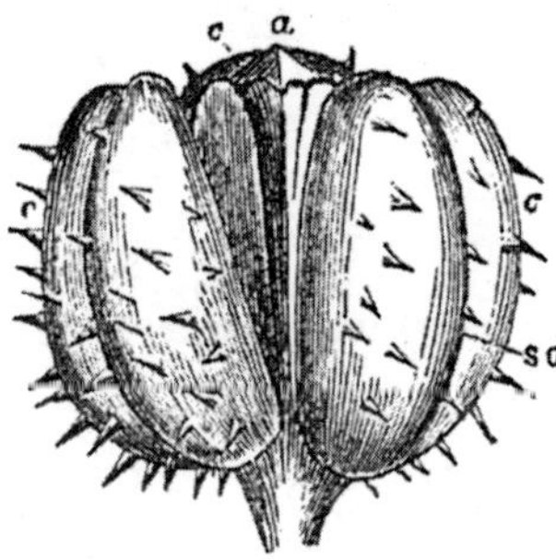

Fig. 215. Ovary of the Castor-oil Plant (*Ricinus communis*), with a portion of the inferior calyx at its base. The styles in this case are once-forked.——*Fig.* 216. Fruit of the Castor-oil Plant dehiscing in a septicidal manner. *c, c, c.* Mature carpels. *a.* Columella. *sd.* Dorsal suture where each carpel ultimately opens.

(*fig.* 218), often caruncalate; *embryo* (*fig.* 218) in fleshy albumen, with flattened cotyledons, and a superior radicle.

Diagnosis.—Herbs, shrubs, or trees, commonly with an acrid milky juice. Flowers unisexual, monœcious or diœcious. Calyx absent or present, and then inferior. Petals rarely present. Male flowers with one or more distinct or united stamens and 2-celled anthers. Female flowers with a superior ovary, 1- or more celled, with 1 or 2 suspended ovules in each cell. Fruit of 1, 2, 3, or many, dry carpels, which separate from the axis and from each other, and usually open with elasticity; or fleshy and indehiscent. Seeds suspended; embryo in fleshy albumen, with flattened cotyledons, and a superior radicle. *Illustra-*

FIG. 217.

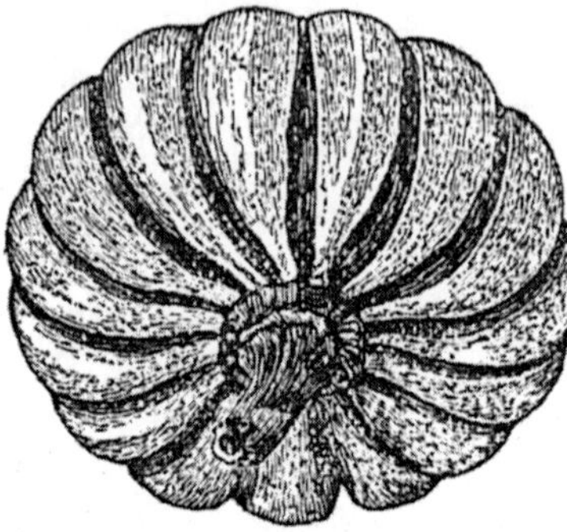

FIG. 218.

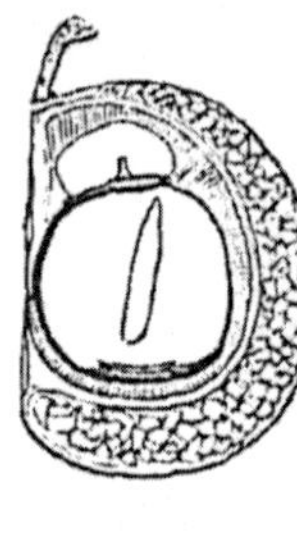

Fig. 217. Fruit of the Sandbox-tree (*Hura crepitans*). It is composed of fifteen carpels which separate from the axis when ripe, and burst with great force.——*Fig.* 218. Vertical section of a coccus of the fruit and seed of a species of *Euphorbia*.

tive Genera:—Euphorbia, *Linn.*; Mercurialis, *Tourn.*; Buxus, *Tourn.*

CUPULIFERÆ, the Oak Order.—Character.—*Trees* or *shrubs*. *Leaves* (*fig.* 219, *b*) alternate, usually feather-veined (*figs.* 220 and 221), simple, with deciduous stipules. *Flowers* monœcious. *Male flowers* (*fig.* 222) clustered or in amenta (*fig.* 223), and with (*fig.* 223) or without bracts; *stamens* (*fig.* 222) 5—20, inserted into the base of a membranous calyx (*fig.* 222), or of scales or bracts (*fig.* 223). *Female flowers* solitary or clustered, and surrounded by an involucre of bracts (*fig.* 224), which ultimately form a cupule round the ovary and fruit (*figs.* 226 and 227); *ovary* inferior, surmounted by a rudimentary calyx (*fig.* 224), 2-, 3- (*fig.* 225), or more celled; *ovules* 2 in each cell (*fig.* 225) or solitary, pendulous or peltate; *stigmas* almost sessile. *Fruit* a glans or nut (*figs.* 226 and 227), 1-celled by abortion, more or less enclosed by

FIG. 219.

FIG. 220

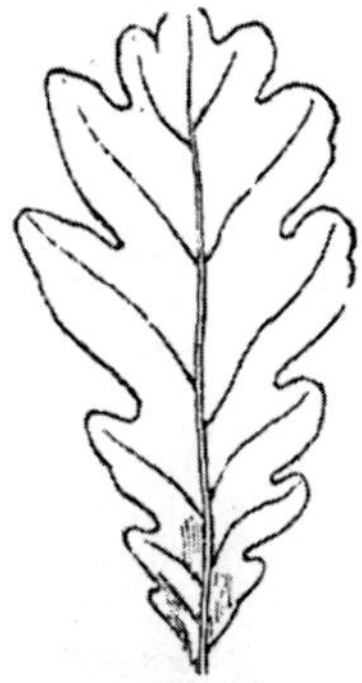

Fig. 219. Branch of the Oak with alternate leaves and leaf-buds in their axils. *a, a.* Buds. *b, b.* Leaves. ——*Fig.* 220. Feather-veined leaf of the Oak.

FIG. 221.

FIG. 222.

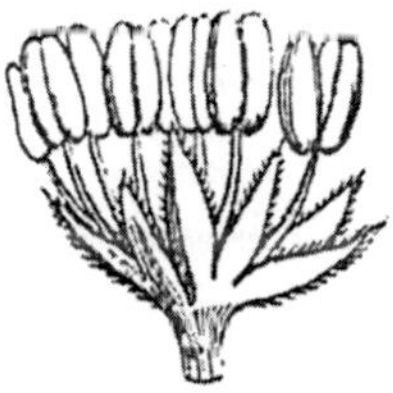

FIG. 223.

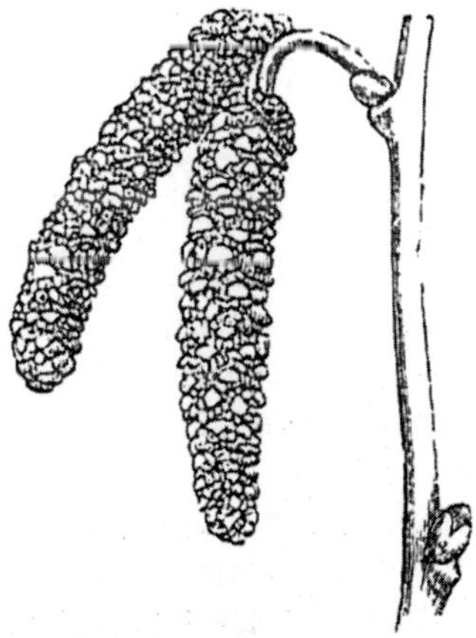

FIG. 224

FIG. 225.

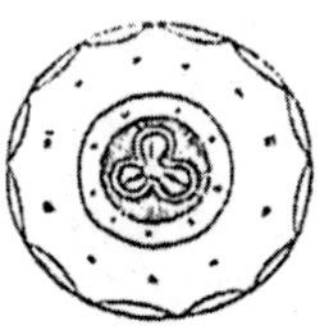

Fig. 221. Feather-veined leaf of the Spanish Chestnut (*Castanea vesca*). —— *Fig.* 222. Male flower of a species of Oak (*Quercus*). —— *Fig.* 223. Staminate or male catkins of the Hazel (*Corylus Avellana*), showing a number of scaly bracts between the flowers.——*Fig.* 224. Female flower of a species of Oak.——*Fig.* 225. Transverse section of the same.

the capsule. *Seeds* large, 1 or 2, without albumen; *cotyledons* thick, fleshy or farinaceous; *radicle* superior.

FIG. 226.

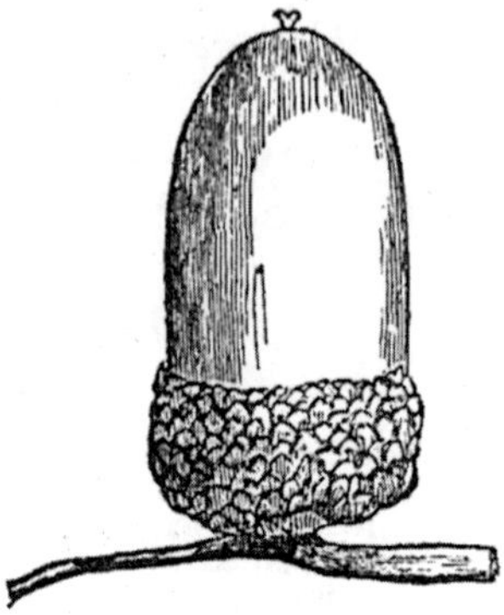

FIG. 227.

Fig. 226. Fruit of the Oak (*Quercus Robur*), surrounded by a cupule. —*Fig.* 227. Fruit of the Hazel (*Corylus Avellana*), with cupule at its base.

Illustrative Genera:—Quercus, *Tourn.*; Carpinus, *Tourn.*; Corylus, *Tourn.*

MYRICACEÆ, the Bog-Myrtle Order.—Character.—*Shrubs* or small *trees*, with alternate simple resinous-dotted leaves, and usually exstipulate. *Flowers* unisexual, amentaceous, monœcious or diœcious, both kinds of flowers in the same or in different catkins. *Male flowers* achlamydeous; *stamens* definite. *Female flowers* achlamydeous, with a 1-celled sessile *ovary*, and 1 erect *ovule*; *fruit* drupaceous; *seed* solitary, erect; *embryo* without albumen; *radicle* superior. *Illustrative Genus:*—Myrica, *Linn.*

BETULACEÆ, the Birch Order.—Character.—*Trees* or *shrubs*. *Leaves* simple, alternate, with deciduous stipules. *Flowers* unisexual, monœcious, amentaceous, with no true *calyx*, but in its place small scaly *bracts*, which in some cases are arranged in a whorled manner.

Male flowers with 2 or 3 *stamens* opposite the bracts. *Female flowers* with a 2-celled *ovary*, and 1 pendulous *ovule* in each cell. *Fruit* dry, indehiscent, often winged, 1-celled, 1-seeded, without a cupule. *Seed* pendulous, exalbuminous; *radicle* superior. *Illustrative Genera*:—Betula, *Tourn.*; Alnus, *Tourn.*

SALICACEÆ, the Willow Order.—Character.—*Trees* or *shrubs. Leaves* (*fig.* 228) simple, alternate, deciduous,

FIG. 228.

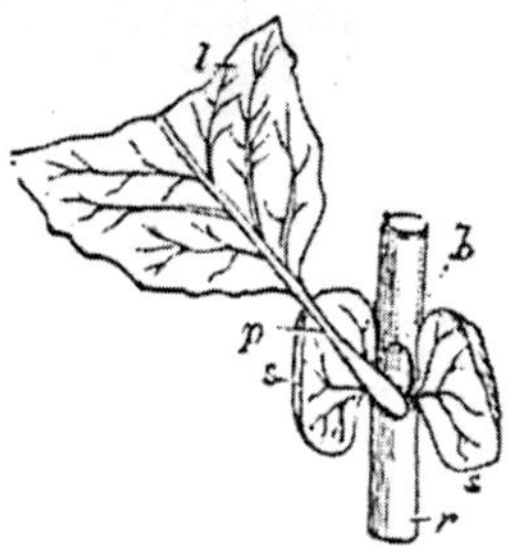

FIG. 229

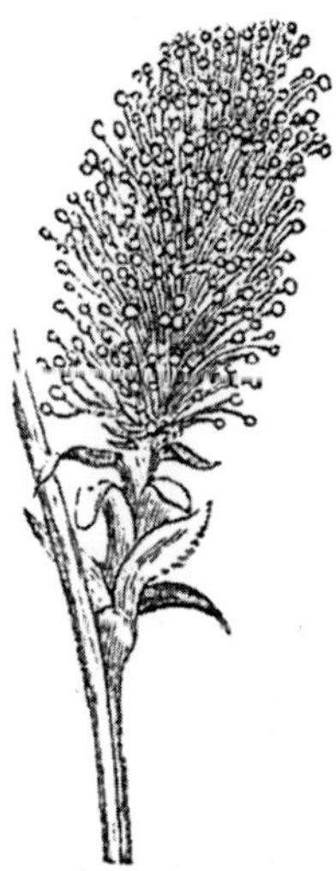

FIG. 230.

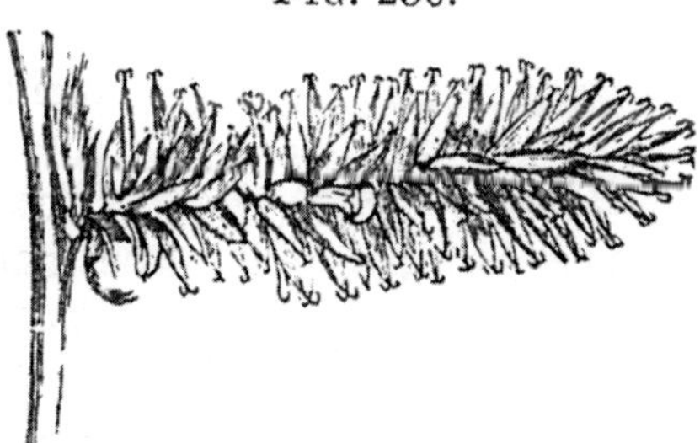

Fig. 228. Leaf and portion of a branch of *Salix aurita*. *r.* Branch. *b.* Bud. *l.* Lamina with the upper portion removed, and attached by a petiole, *p*, to the stem. *s, s.* Caulinary stipules. ——*Fig.* 229. Staminate flowers of a species of Willow (*Salix*). ——*Fig.* 230. Pistillate or carpellary flowers of the same plant.

with deciduous or persistent stipules. *Flowers* unisexual, diœcious, amentaceous (*figs.* 229 and 230), naked, or with a membranous or cuplike *calyx*. *Male flowers* (*figs.* 229 and 231) with 1—30 distinct or monadelphous stamens. *Female flowers* sessile or stalked (*fig.* 232), with a 1-celled *ovary*; 1 style (*fig.* 232), 2 (*fig.* 232) or 4 stigmas, and numerous erect *ovules*. *Fruit* 1-celled, dehiscing loculicidally, 2-valved. *Seeds* numerous, with long silky hairs (*fig.* 233) springing from a short funiculus and covering the seed, exalbuminous; *embryo* erect, with an inferior radicle. *Illustrative Genera* :—Salix, *Tourn.*; Populus, *Tourn.*

AMENTACEÆ.—The four preceding orders—namely, Cupuliferæ, Myricaceæ, Betulaceæ, and Salicaceæ, belong to the alliances *Quernales* and *Amentales* of Lindley. The

FIG. 231.

FIG. 232.

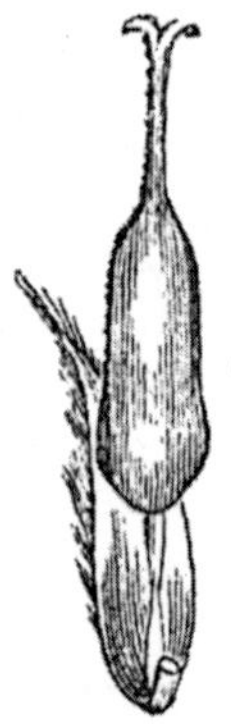

FIG. 233.

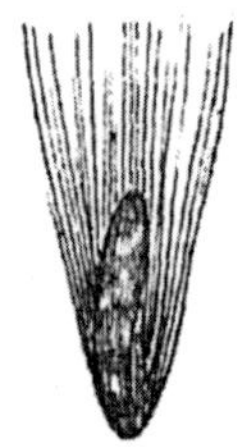

Fig. 231. Male flower of a species of Willow (*Salix*), with two stamens, and a single bract at the base.——*Fig.* 232. Female flower of the same with bract at the base, and a solitary stalked ovary and style surmounted by two stigmas.——*Fig.* 233. Hairy seed of a species of Willow (*Salix*).

order Cupuliferæ belongs to the former, and the Myricaceæ, Betulaceæ, and Salicaceæ to the latter; and they together constitute the Amentaceæ of the Syllabus of the University of London. The two alliances may be characterised as follows :—

Alliance *Quernales*.—Flowers unisexual in amenta, or clustered, or solitary; male flowers monochlamydeous; carpels inferior; embryo amygdaloid, without albumen.

Alliance *Amentales*.—Flowers unisexual, all in amenta; achlamydeous or monochlamydeous; carpels superior; embryo small, with little or no albumen.

The orders may also be simply arranged and characterised as follows:—

1. Ovary inferior.

Cupuliferæ.—Ovary 2- or more celled. Ovules pendulous or peltate. Fruit with a cupule, 1-celled.

2. Ovary superior.

Myricaceæ.—Ovary 1-celled. Ovule solitary, erect. Fruit drupaceous, indehiscent, without a cupule. Seed solitary, without hairs.

Betulaceæ.—Ovary 2-celled. Ovule solitary in each cell, pendulous. Fruit thin, without a cupule, indehiscent, often winged, 1-celled. Seed solitary, without hairs.

Salicaceæ.—Ovary 1-celled. Ovules numerous, erect. Fruit 1-celled, dehiscing loculicidally, 2-valved. Seeds numerous, with cottony hairs.

CLASS I. DICOTYLEDONES.

Division II. GYMNOSPERMIA.

CONIFERÆ, the Coniferous Order.—Character.—Resinous *trees* or evergreen *shrubs*, with branched

FIG. 234.

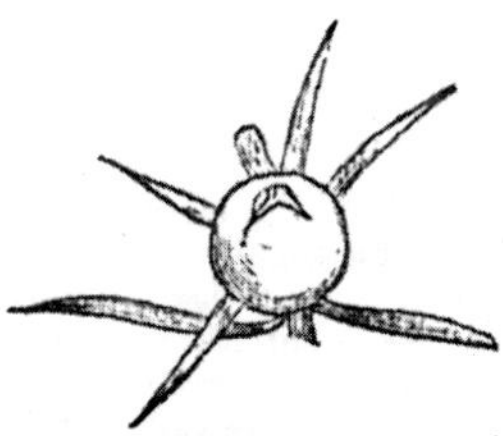

FIG. 235.

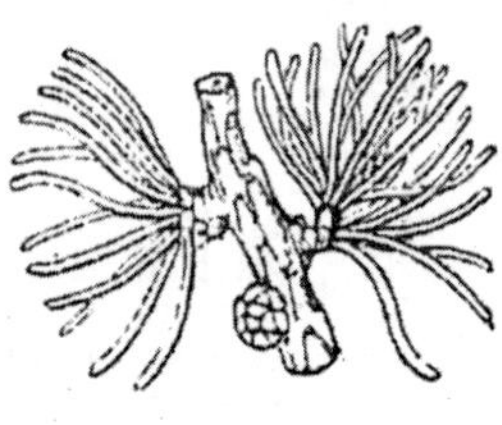

Fig. 234. Acicular leaves of Juniper (*Juniperus communis*).——*Fig.* 235. Fascicled or tufted linear leaves of the Larch (*Abies* or *Pinus Larix*, or *Larix europæa*).

stems. *Leaves* linear (*fig.* 235), acicular (*fig.* 234), or lanceolate, parallel-veined, fascicled (*fig.* 235) or imbricate. *Flowers* naked, monœcious or diœcious. *Male flowers* arranged in deciduous amenta. *Stamens* 1 or several, in the latter case monadelphous; *anthers*

1- or more celled, opening longitudinally. *Female flowers* in cones (*fig.* 236), consisting of flattened imbricated carpels or scales arising from the axil of membranous bracts; *ovules* naked, 2 (*fig.* 237, *ov*) or more, on the upper surface of each carpel. *Fruit* a woody cone (*figs.* 238 and 239) or a galbulus (*figs.* 240 and

FIG. 236.

FIG. 237.

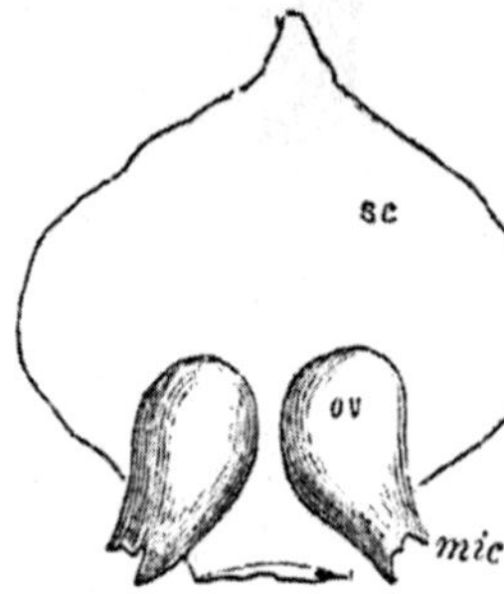

FIG. 238.

FIG. 239.

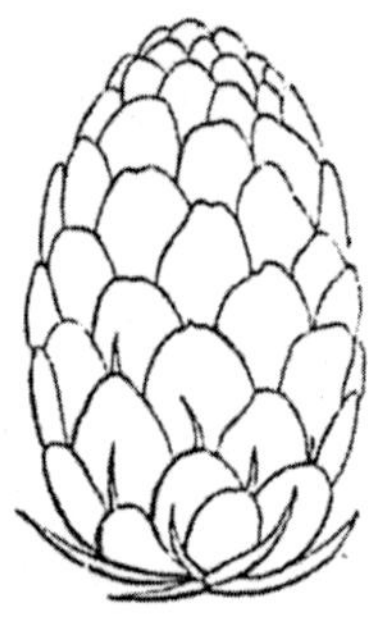

FIG. 240

Fig. 236. Cone of Hemlock Spruce (*Pinus* or *Abies canadensis*).——*Fig.* 237. Bract or carpellary leaf, *sc*, of a species of *Pinus*, bearing two naked ovules, *ov*, at its base. *mic.* Micropyle.——*Fig.* 238. Cone or fruit of the Scotch Fir.——*Fig.* 239. A ripe cone of the Larch (*Abies* or *Pinus Larix*, or *Larix europæa*).——*Fig.* 240. Galbulus or fruit of the Juniper (*Juniperus communis*).

241). *Seeds* naked (*figs.* 242 and 243), with a hard crustaceous integument, often winged (*figs.* 243 and 244, *w*), albuminous; *cotyledons* 2 or many (*fig.* 245).

Division of the Order and Illustrative Genera:—This order has been subdivided as follows:—

Sub-order 1. *Abieteæ.*—Ovules inverted, with the

micropyle next the base of the carpel (*fig.* 237). Pollen oval. *Illustrative Genus* :—Pinus, *Linn.*

FIG. 241.

FIG. 243.

FIG. 242.

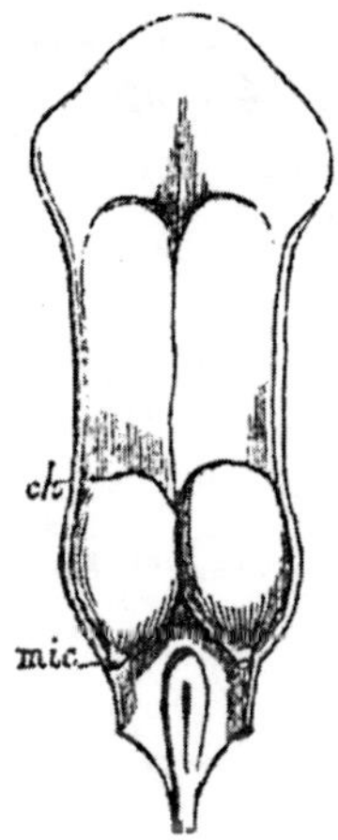

FIG 244.

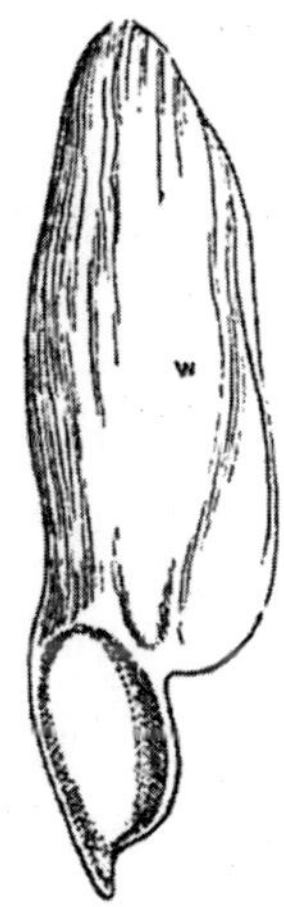

FIG. 245.

Fig. 241. Galbulus or fruit of the Cypress (*Cupressus sempervirens* ——*Fig.* 242. A mature carpel or scale of the Scotch Fir (*Pinu sylvestris*), with two winged naked seeds at its base. *mic.* Micropyle. *ch.* Chalaza.——*Fig.* 243. A scale of the Larch bearing one naked seed; the other seed has been removed.——*Fig.* 244. Seed of a species of *Pinus*, with a winged appendage, *w.*——*Fig.* 245. The so-called polycotyledonous embryo of a species of *Pinus* beginning to germinate. *c.* Cotyledons. *r.* Radicle. *t.* Tigellum.

Sub-order 2. *Cupresseæ.*—Ovules erect, with micropyle superior. Pollen spheroidal. *Illustrative Genus* :—Juniperus, *Linn.*

CLASS II. MONOCOTYLEDONES.

Sub-class I. *Petaloideæ.*

1. Epigynæ.

ORCHIDACEÆ, the Orchis Order.—Character.—*Herbs* or *shrubs*, terrestrial (*figs.* 246 and 247) or epiphytic (*fig.* 248). *Roots* fibrous (*fig.* 248, *a*, *a*) or tuberculated (*figs.* 246 and 247); no true stem or a pseudobulb (*fig.* 248, *b*). *Leaves* entire (*fig.* 249),

FIG. 246.

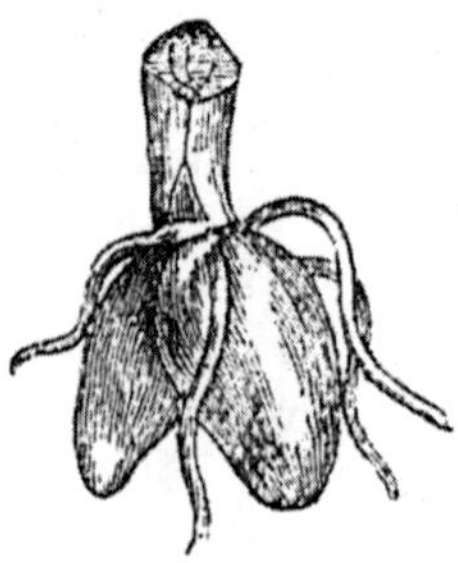

FIG. 247.

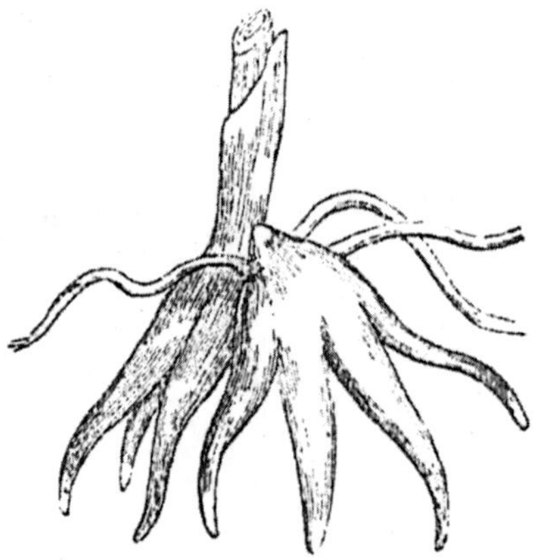

Fig. 246. Tubercular roots of an Orchis.—*Fig.* 247. Palmated tubercules of an Orchis.

generally sheathing. *Flowers* irregular (*figs.* 250—252), solitary or numerous, with a single bract (*fig.* 250), hermaphrodite (*fig.* 251). *Perianth* superior (*fig.* 250), usually petaloid and composed of six pieces (*fig.* 251), which are commonly arranged in two whorls; the *outer whorl*, *s*, *sl*, *sl*, formed of three pieces (*sepals*), more or less united below or distinct; one, *s*, being anterior or inferior, or when the ovary is twisted posterior (*figs.* 250 and 252).

FIG. 248.

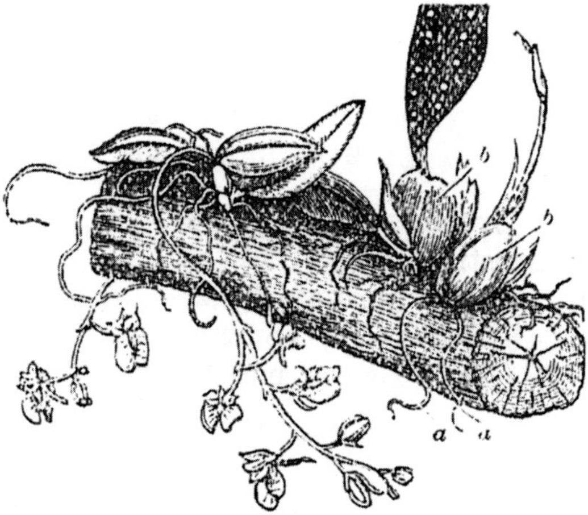

FIG. 249.

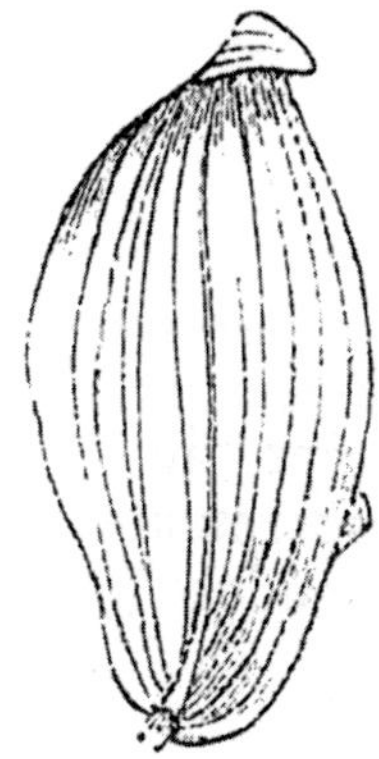

Fig. 248. Epiphytic Orchid. *a*, *a*. Aerial roots. *b*, *b*. Pseudobulbs. ——*Fig.* 249. Leaf of an Orchis showing parallel venation; the margins are entire.

FIG. 250.

FIG. 251.

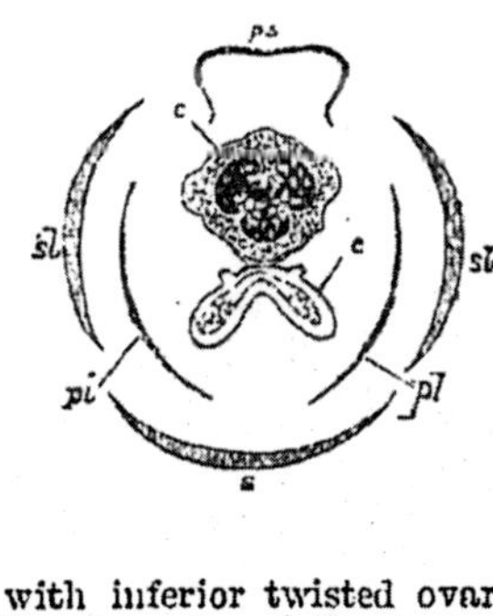

Fig. 250. Flower of *Orchis mascula*, with inferior twisted ovary. The column in the centre is formed by the union of the stamens and style. The labellum or lip at the anterior part of the flower is spurred. Below the flower a solitary bract may be seen.——*Fig.* 251. Diagram of the flower of an Orchid. *s*, *sl*, *sl*. The three outer divisions of the perianth; *s* being anterior or inferior, *sl*, *sl* being lateral. *pl*, *pi*. The two lateral divisions of the inner whorl of the perianth. *ps*. The superior or posterior division (*labellum*) of the inner whorl; this by the twisting of the ovary becomes ultimately inferior or anterior. *e*. The fertile stamen, with two anther lobes. *c*. Transverse section of the ovary, with three parietal placentas.

and two *sl*, *sl*, lateral; the *inner whorl* (*fig.* 251, *pi*, *pl*, *ps*), usually consists of three pieces (*petals*), (or rarely of but one), alternating with the pieces in the outer whorl; one (the *labellum* or *lip*) (*fig.* 251, *ps*) posterior or superior, or by the twisting of the ovary anterior (*figs.* 250 and 252), usually longer and larger than the other pieces, and altogether different to them in form (*fig.* 252), often spurred (*fig.* 250); some-

FIG. 252.

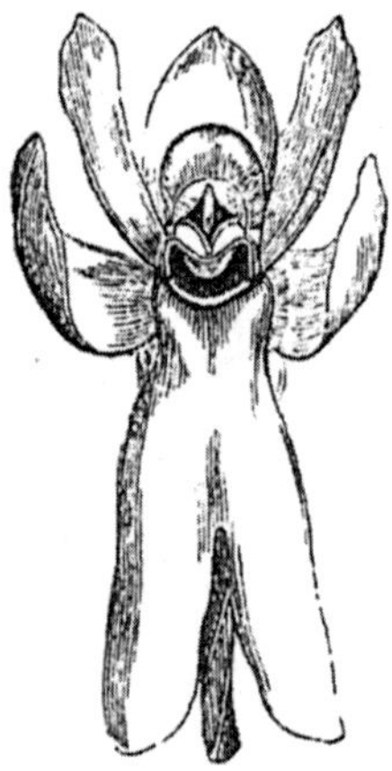

FIG. 253.

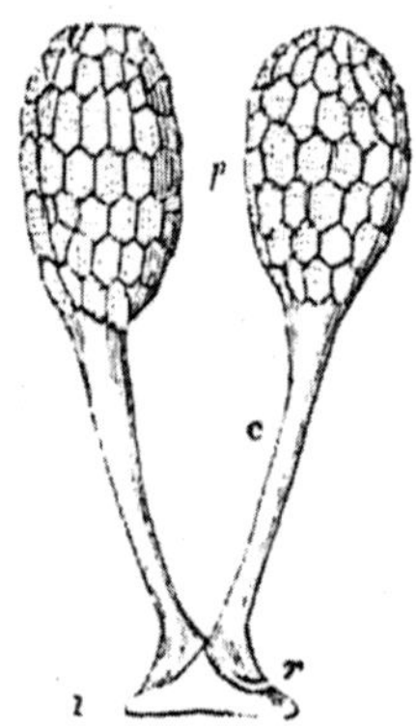

Fig. 252. Front view of the flower of the Tway-blade (*Listera ovata*), showing the bifid labellum at the anterior part of the flower, and the other five divisions of the perianth; and the essential organs of reproduction forming a column (*gynostemium*).——*Fig.* 253. Pollinia, *p*, of Orchis with their caudicles, *c*, and the retinacula, *r*, *r*, of the rostellum of the stigma, at the base.

times the labellum exhibits a division into three regions of which the lowest is then termed the *hypochilium*, the middle the *mesochilium*, and the upper the *epichilium*. *Andrœcium* united to the style (*gynandrous*) (*figs.* 250—252) and forming with it a central column (*gynostemium*); the *column* bearing 1 perfect anther (*fig.* 251, *e*) and two lateral abortive ones, or, in *Cypripedium*, two lateral perfect anthers and one abortive anther in the centre. *Pollen* powdery, or more or less collected into grains or waxy or mealy masses (*pollinia*) (*fig.* 253, *p*); the masses free or attached by their stalk *c* (*caudicle*) to a gland or glands (*retinacula*) at the apex (*rostellum*) of the stigma (*fig.* 254, *a*). *Ovary* inferior (*fig.* 250), 1-celled, with 3 parietal placentas (*figs.* 251, *c*, and 255, *pl*) bearing a number of anatropous ovules; *style* united with the andrœcium and forming with it a column or *gynostemium* (*figs.*

FIG. 254.

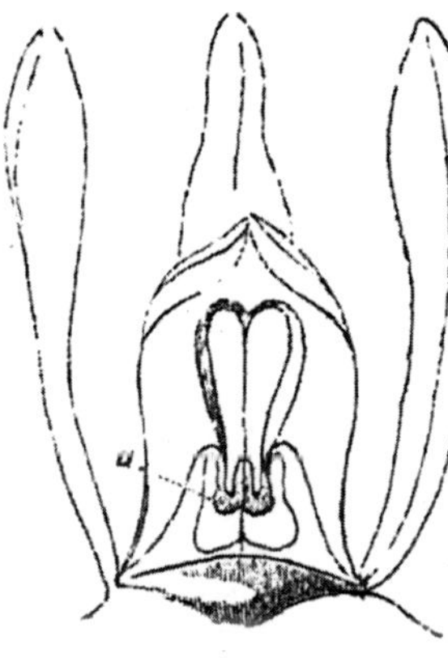

FIG. 255.

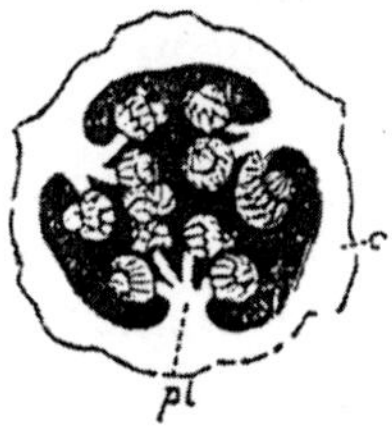

FIG. 256.

FIG. 257.

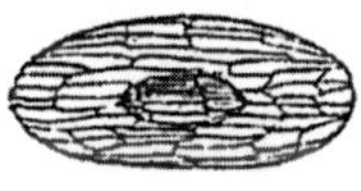

Fig. 254. Upper part of the flower of an Orchis, showing the pollinia adhering to the stigma by the retinacula, *a*.——*Fig.* 255. Transverse section of the one-celled ovary of an Orchis. *c*. One of the three carpels of which it is formed, slightly infolded at its margins. *pl*. One of the three parietal placentas.——*Fig.* 256. Fruit of an Orchid dehiscing by three valves, each valve bearing a placenta and numerous very minute seeds.——*Fig.* 257. Seed of an Orchid, with a loose reticulated testa.

250—252); *stigma* a viscid space in front of the column (*fig.* 254). *Fruit* usually capsular, 3-valved (*fig.* 256), the valves bear the placentas in their middle, and separating when the fruit is ripe from the central parts or midribs of the component carpels, which are left as an open framework; or rarely fleshy and indehiscent. *Seeds* very minute and numerous, with a loose netted (*fig.* 257) or rarely hard crustaceous testa, exalbuminous; *embryo* a fleshy solid mass.

Diagnosis.—This order is readily known by its irregular flowers; by the peculiar form which the labellum in many cases assumes, so as to cause the flower to resemble some insect, reptile, bird, or other living object; by its gynandrous stamens; by its frequently more or less coherent pollen; and by its 1-celled inferior ovary with three parietal placentas.

Division of the Order and Illustrative Genera:—This order has been divided by Lindley and others into several tribes, the characters being derived from the number and position of the anthers, the number and nature of the pollen-masses, &c. These need, however, no explanation in this volume. *Illustrative Genera:*—Orchis, *Linn.*; Ophrys, *Linn.*; Epipactis, *Rich.*; Epipogum, *Gmelin*; Malaxis, *Swartz*; Cypripedium, *Linn.*

IRIDACEÆ, the Iris Order.—Character.—*Herbs,*

FIG. 258.

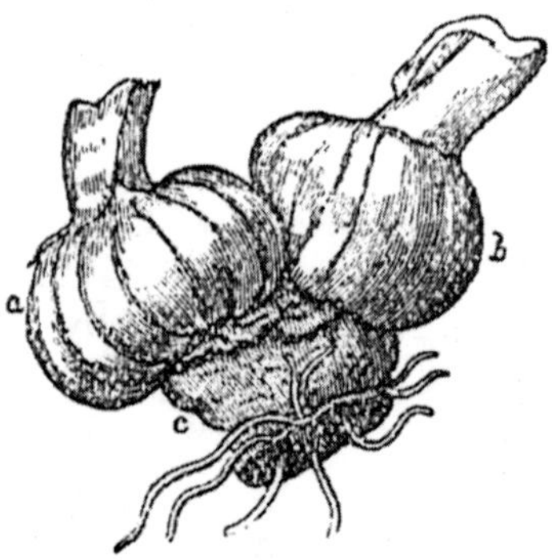

FIG. 259.

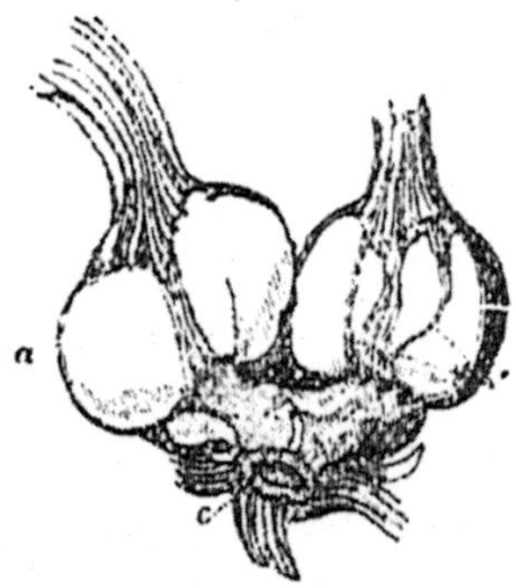

Fig. 258. Corms of the Saffron plant (*Crocus sativus*). *a*, *b*. The new corms, arising from *c*, the apex of the old or parent corm. —*Fig.* 259. Vertical section of the former. The letters refer to the same parts in the two figures.

usually with bulbs, corms (*figs.* 258 and 259), or rhizomes (*fig.* 260). *Leaves* with parallel venation, generally equitant (*fig.* 260). *Flowers* spathaceous. *Perianth* superior (*fig.* 261, *t*), petaloid, 6-partite (*figs.*

262 and 263), in 2 whorls, which are equal or nearly so (*fig*. 263), or unequal (*figs*. 261 and 262) in the size of their segments; or sometimes the parts are entirely distinct. *Stamens* 3, inserted on the outer segments of the perianth (*fig*. 263); *anthers* 2-celled, innate, extrorse (*fig*. 263). *Ovary* inferior (*fig*. 261, *o*),

FIG. 260.

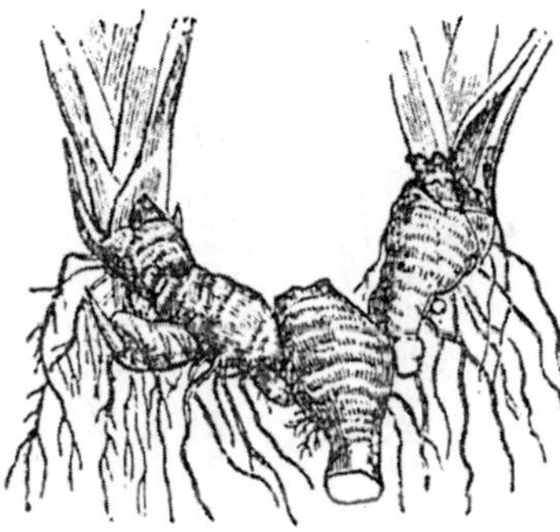

FIG. 261.

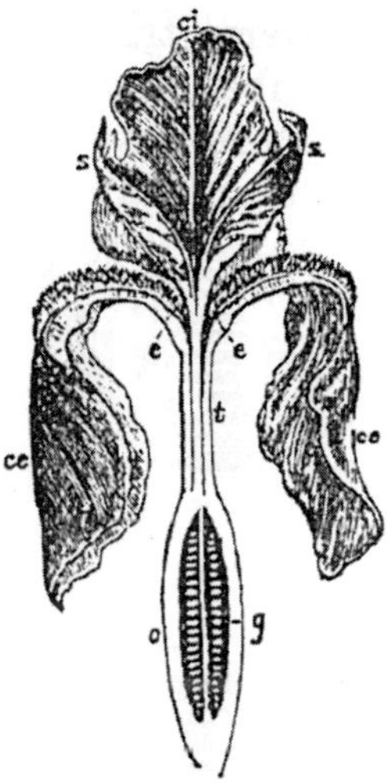

Fig. 260. A portion of the rhizome of a species of *Iris*. The bases of the leaves are equitant.——*Fig*. 261. Vertical section of the flower of *Iris germanica*. *ce*, *ce*. Two of the external and larger divisions of the perianth. *ci*. One of the internal smaller divisions. *t*. Tube formed by the union of the divisions of the perianth. *e*, *e*. Stamens, covered by the petaloid stigmas, *s*, *s*. *o*. Inferior ovary, with numerous ovules, *g*, attached to placentas in the axis.

3-celled, with axile placentation (*fig*. 262); *style* 1 (*figs*. 263 and 264); *stigmas* 3, often petaloid (*figs*. 264 and 265, *stig*). *Fruit* capsular, 3-celled, 3-valved, with loculicidal dehiscence (*fig*. 266). *Seeds* numerous, with horny or fleshy albumen (*fig*. 267).

Diagnosis.—Herbs. Leaves with parallel veins. Flowers on scapes and spathaceous. Perianth petaloid, superior, 6-partite or rarely 6-leaved, in 2 equal or unequal whorls. Stamens 3; anthers innate, extrorse.

FIG. 262.

FIG. 263.

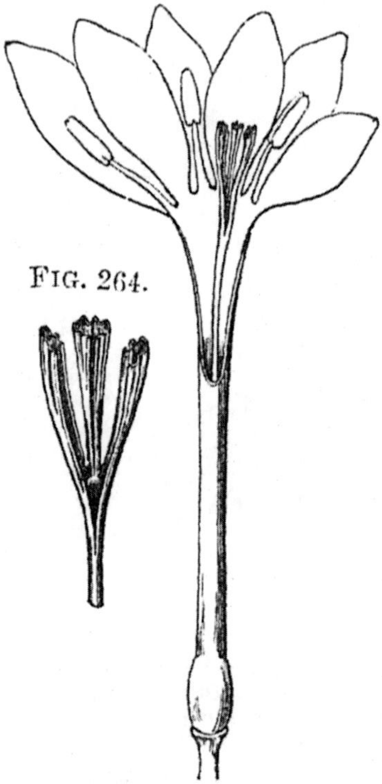

FIG. 265.

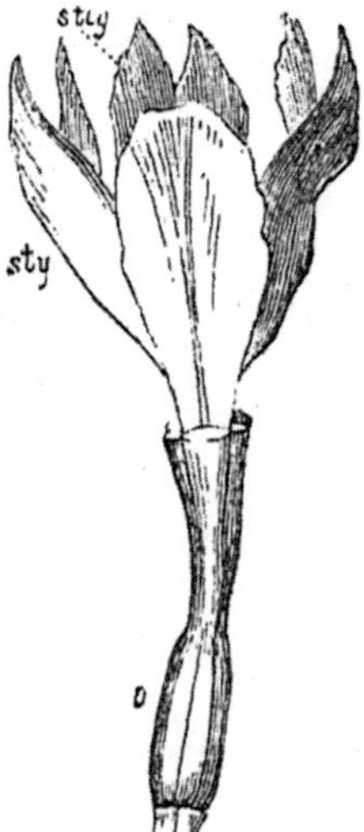

FIG. 264.

Fig. 262. Diagram of the flower of a species of *Iris*, showing a solitary bract below, six divisions to the perianth arranged in two whorls, three stamens, and a three-celled ovary, with axile placentas.——*Fig.* 263. A flower of the Spring Crocus (*Crocus vernus*) cut open to show the three extrorse stamens attached to the outer segments of the perianth.——*Fig.* 264. The three petaloid stigmas of the same with the end of the style.——*Fig.* 265. Pistil of a species of *Iris*. *o*. Inferior ovary. *sty*. Style and three petaloid stigmas.

Ovary 3-celled, with axile placentas, inferior. Fruit capsular, with loculicidal dehiscence, 3-celled, 3-valved.

Seeds numerous, albuminous. *Illustrative Genera:*— Iris, *Linn.*; Crocus, *Linn.*

FIG. 266.

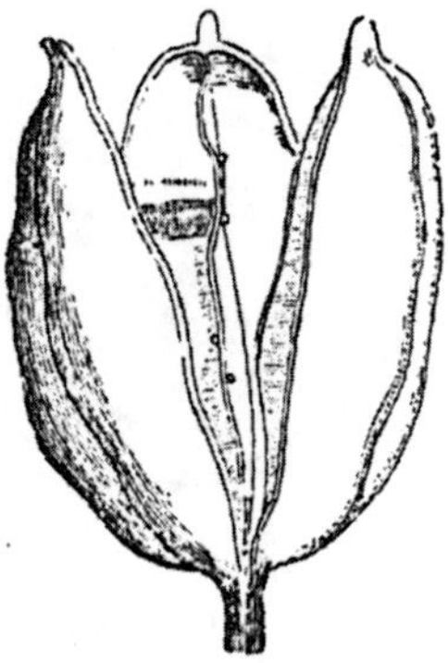

FIG. 267.

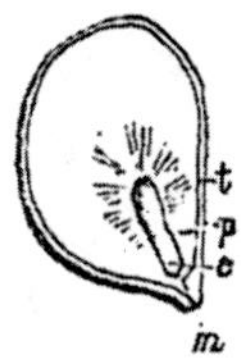

Fig. 266. Capsular fruit (*diplotegia*) of a species of Iris, opening in a loculicidal manner.—*Fig.* 267. Vertical section of the seed of *Iris germanica*. *t.* Integuments of the seed. *p.* Albumen. *e.* Embryo. *m.* Micropyle.

AMARYLLIDACEÆ, the Amaryllis Order.—Bulbous or fibrous-rooted plants, without any aerial stem, or sometimes with a woody one. *Leaves* with parallel venation, and usually linear-ensiform. *Flowers* generally on scapes and spathaceous (*fig.* 268). *Perianth* regular or nearly so (*figs.* 268 and 270), petaloid, superior (*fig.* 272), with six divisions and with (*figs.* 270 and 271, *n*), or without, a corona (*fig.* 272). *Stamens* 6, inserted on the perianth (*fig.* 271) or summit of the ovary (*fig.* 272); *anthers* 2-celled, introrse (*fig.* 272). *Ovary* inferior (*fig.* 272), 3-celled (*fig.* 269); *placentas* axile (*figs.* 269 and 272). *Fruit* capsular, 3-celled, 3-valved, with loculicidal dehiscence, and numerous seeds; or a berry with 1—3 seeds. *Seeds* with fleshy or horny albumen; *embryo* with the radicle next the hilum (*fig.* 273).

Diagnosis.—Leaves with parallel veins. Flowers

spathaceous. Perianth superior, petaloid, regular, 6-partite, frequently with a corona. Stamens 6; anthers

FIG. 269.

FIG. 268.

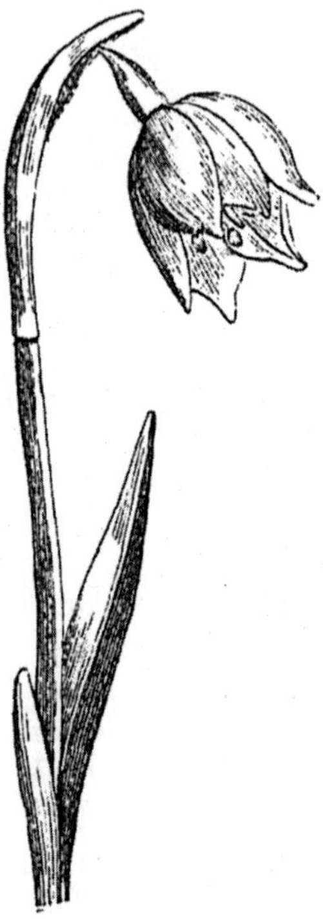

FIG. 270.

Fig. 268. Flower of the Spring Snowflake (*Leucojum vernum*), with spathaceous bract.——*Fig.* 269. Diagram of the same, with six divisions to the perianth arranged in two whorls, six stamens, and a 3-celled ovary with axile placentation.——*Fig.* 270. Flower of the Daffodil (*Narcissus Pseudo-narcissus*). The bell-shaped process towards the centre is termed a corona.

introrse. Ovary inferior, 3-celled, with axile placentas. Fruit capsular, 3-valved, with loculicidal dehiscence, or baccate. Seeds numerous, albuminous. *Illustrative Genera* :—Narcissus, *Linn.*; Galanthus, *Linn.*

FIG. 271.

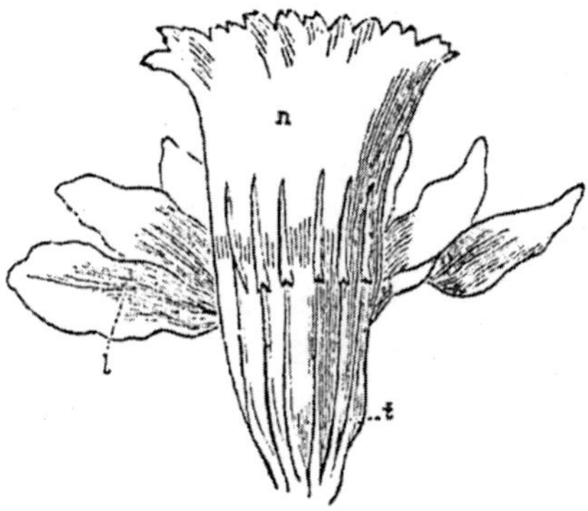

Fig. 271. The perianth of the Daffodil (*Narcissus Pseudo-narcissus*) cut open in a vertical manner. *t.* Tube bearing six stamens. *l.* Limb of the perianth. *n.* Corona.

FIG. 272.

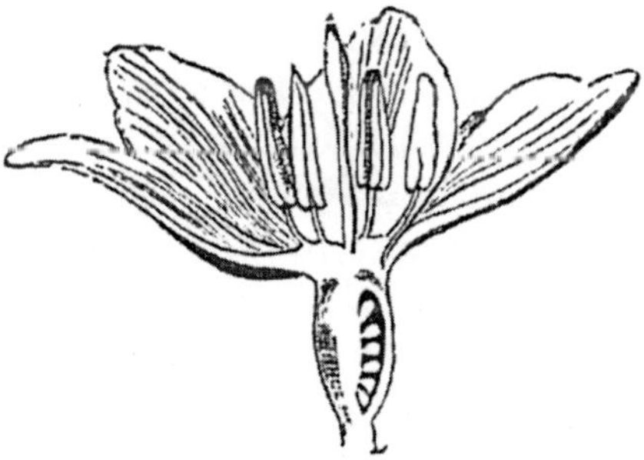

FIG. 273.

Fig. 272. Vertical section of the flower of the Spring Snowflake (*Leucojum vernum*), showing inferior ovary. The placentation is axile. ——*Fig.* 273. Vertical section of the seed of the same.

2. Hypogynæ.

LILIACEÆ, the Lily Order.—Character.—*Herbs* (*figs.* 276 and 277), *shrubs* (*fig.* 274), or *trees* (*fig.* 275), with bulbs (*figs.* 276 and 277), rhizomes (*fig.* 278), tubers, or fibrous roots. *Stem* simple or branched (*fig.* 275). *Leaves* with parallel veins, sessile or sheathing, sometimes succulent (*fig.* 279, *l*). *Flowers* regular (*figs.* 280 and 281), variously arranged (*figs.* 274, 279 and 282). *Perianth* green or petaloid, inferior (*figs.* 280 and 283), 6-leaved (*fig.* 280) or 6-partite (*fig.* 279). *Stamens* 6 (*figs.* 280, 281 and 284), or 3 in *Ruscus*, inserted on the perianth (*fig.* 283) or rarely on the thalamus (*fig.* 284); *anthers* introrse (*fig.* 284). *Ovary* superior (*figs.* 283 and 284), 3-celled, with numerous ovules on axile placentas (*figs.* 281 and 285); *style* 1 (*fig.* 286) or absent (*fig.* 284); *stigma* simple (*figs.* 280 and 283) or 3-lobed (*fig.* 286). *Fruit* a loculicidal capsule, or succulent and indehiscent, 3-celled. *Seeds* with fleshy albumen (*fig.* 287), numerous.

FIG. 274.

Fig. 274. Portion of a branch of the Butcher's Broom (*Ruscus aculeatus*), with phylloid pedicels (*cladodes*) bearing flowers, *a*.

Diagnosis.—Leaves with parallel veins, or succulent. Flowers regular. Perianth inferior, 6-leaved or 6-cleft. Stamens 6, or 3 in *Ruscus*; anthers introrse. Ovary superior, 3-celled, with axile placentation; style 1, undivided or absent; stigma simple or 3-lobed. Fruit indehiscent or a loculicidal capsule. Seeds numerous, albuminous.

Division of the Order and Illustrative Genera:—This order has been divided by Baker into three tribes as follows:—

Tribe 1. *Lilieæ.*—Anthers introrse (*fig.* 284). Styles united (*fig.* 286). Fruit a loculicidal capsule. *Illustrative Genera*:—Lilium, *Linn.*; Tulipa, *Linn.*; Scilla, *Linn.*

Tribe 2. *Colchiceæ.*—Anthers extrorse. Styles separate (*fig.* 289). Fruit a septicidal capsule (*fig.* 292).

(This tribe forms the natural order Melanthaceæ or Colchicaceæ of this volume.)

FIG. 275.

Fig. 275. The Dragon Tree of Teneriffe (*Dracæna Draco*), now destroyed.

FIG. 276.

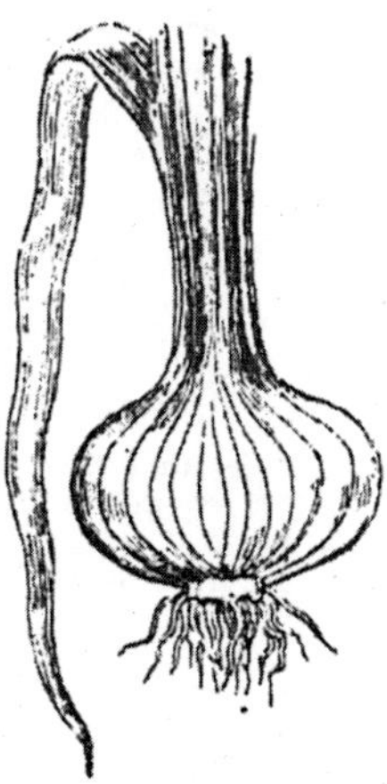

FIG. 277.

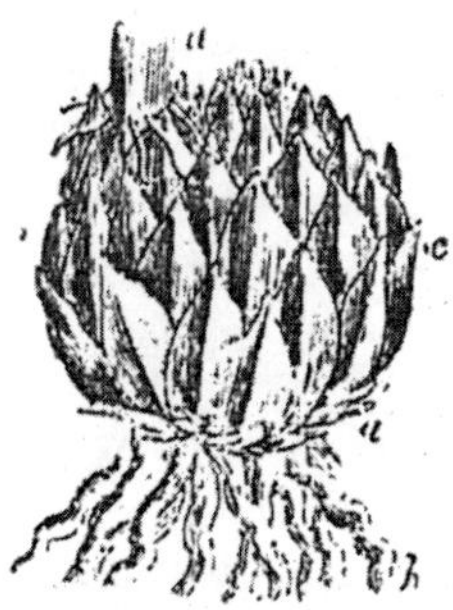

Fig. 276. Tunicated bulb of the Onion (*Allium Cepa*).——*Fig.* 277. Scaly bulb of a species of Lily (*Lilium*). *a.* Shortened axis. *b.* Roots. *c.* Scales. *d.* Flowering stem.

FIG. 278.

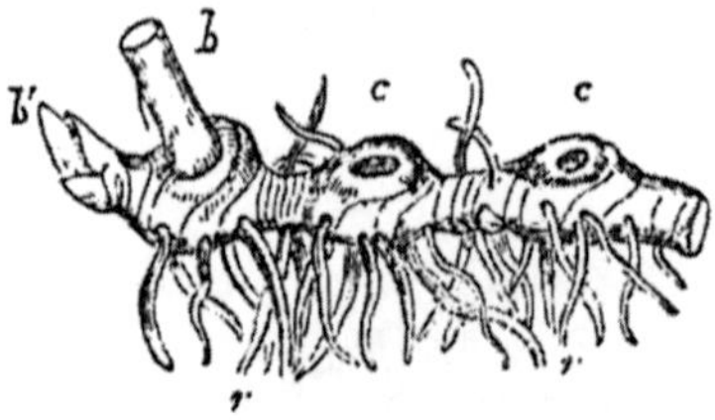

FIG. 279.

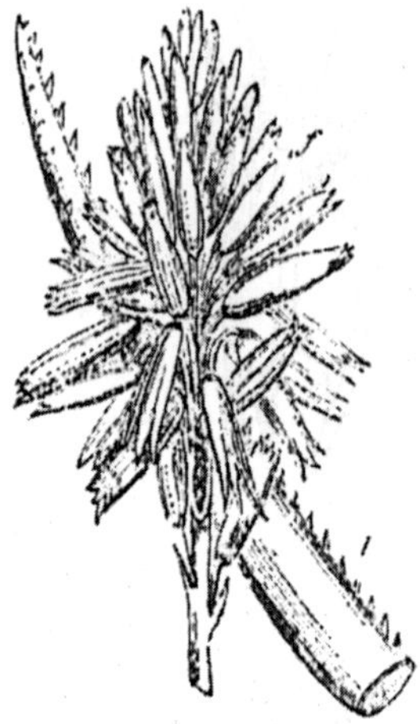

FIG. 280.

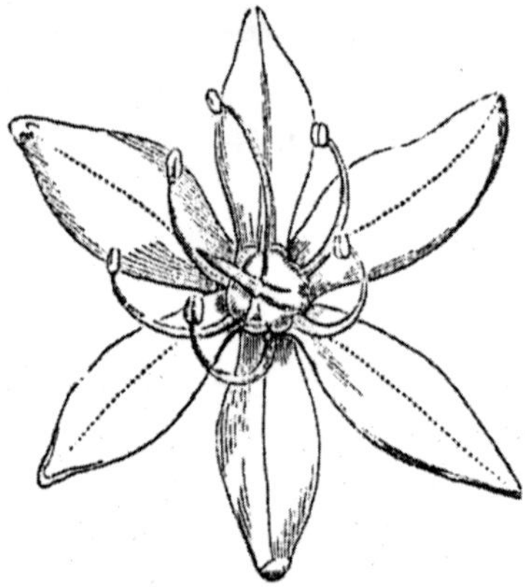

FIG. 281.

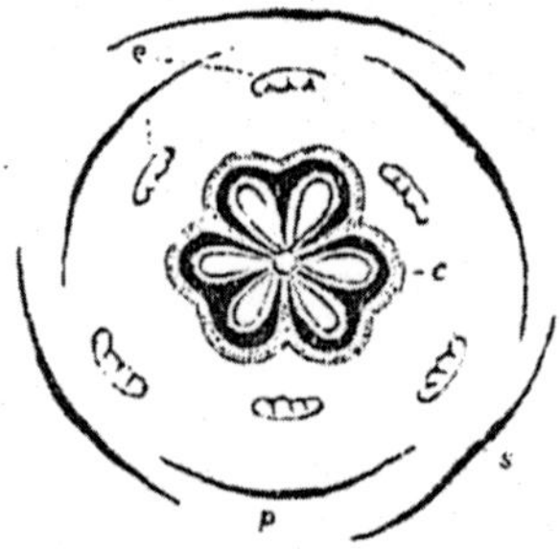

FIG. 282.

Fig. 278. A portion of the rhizome of the Solomon's Seal (*Polygonatum multiflorum*). *b*. Remains of former aerial branch. *b'*. Terminal bud. *c*, *c*. Scars produced by the decay of old branches. *r*, *r*. Roots.——*Fig.* 279. Raceme of flowers, *f*, and portion of the succulent leaf, *l*, of a species of *Aloe*.——*Fig.* 280. Flower of a species of Squill (*Scilla italica*). The parts composing the floral envelopes here resemble one another, and form collectively a perianth.——*Fig.* 281. Diagram of the flower of a species of Lily (*Lilium*). *s*. The three outer divisions of the perianth. *p*. The three inner. *e*. The six stamens. *c*. Three-celled ovary.——*Fig.* 282. Simple umbel of a species of *Allium*.

Tribe 3. *Asparageæ.*—Fruit baccate. *Illustrative Genera:*—Asparagus, *Linn.*; Convallaria, *Linn.*

FIG. 283.

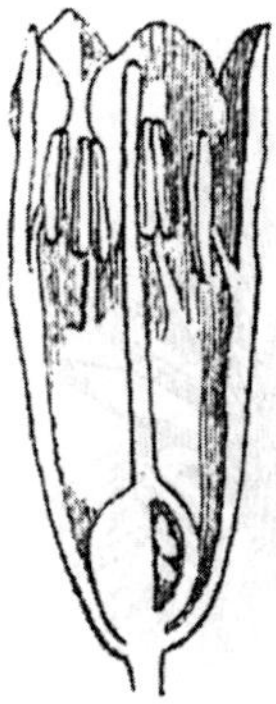

FIG. 284.

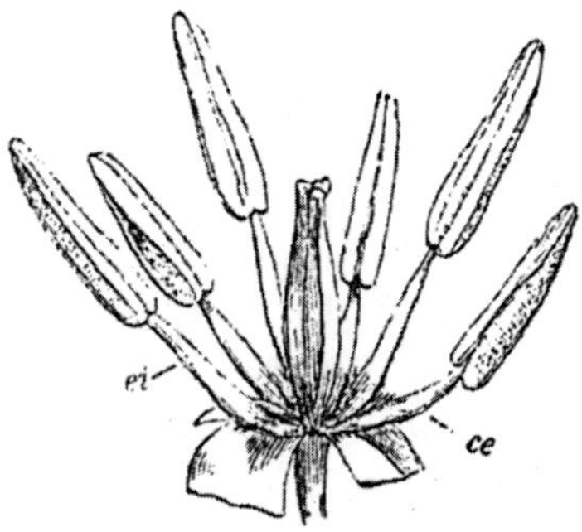

FIG. 286.

FIG. 285.

FIG. 287.

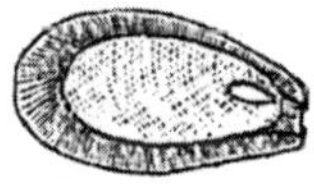

Fig. 283. Vertical section of the flower of the Solomon's Seal (*Polygonatum multiflorum*).——*Fig.* 284. Gynœcium and andrœcium of a species of Tulip (*Tulipa*). The stamens are hypogynous, and the anthers are introrse, and dehisce longitudinally.——*Fig.* 285. Transverse section of the ovary of the White Lily (*Lilium candidum*).——*Fig.* 286. Pistil of a species of Lily (*Lilium*), with one style and a trilobate stigma.——*Fig.* 287. Vertical section of the seed of the Crown Imperial (*Fritillaria imperialis*).

MELANTHACEÆ OR COLCHICACEÆ, the Colchicum Order.—Character.—*Herbs*, with bulbs, rhizomes, corms (*figs.* 288 and 289), tubers, or fibrous roots. *Flowers* regular (*figs.* 289 and 290), usually hermaphrodite (*fig.* 290) or rarely unisexual. *Perianth* inferior, white, green, or purple, 6-partite (*figs.* 289—291) or 6-leaved. *Stamens* 6 (*figs.* 290 and 291); *anthers* extrorse (*fig.*

291). *Ovary* superior or nearly so, 3-celled with axile placentation (*fig.* 290); *style* 3-partite (*fig.* 289); *stigmas* 3. *Fruit* 3-celled (*fig.* 293), 3-valved, with

FIG. 288.

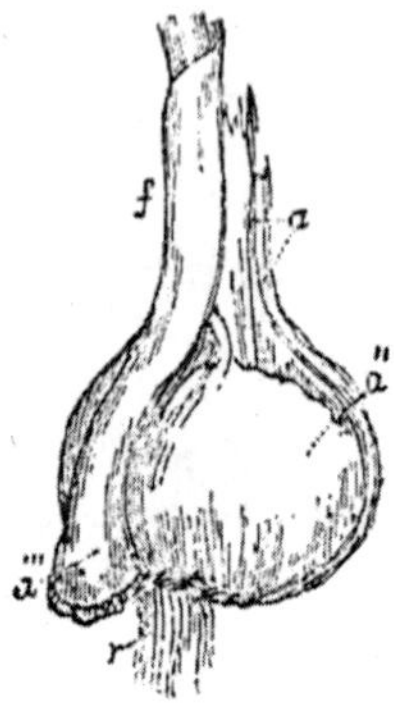

FIG. 289.

FIG. 290.

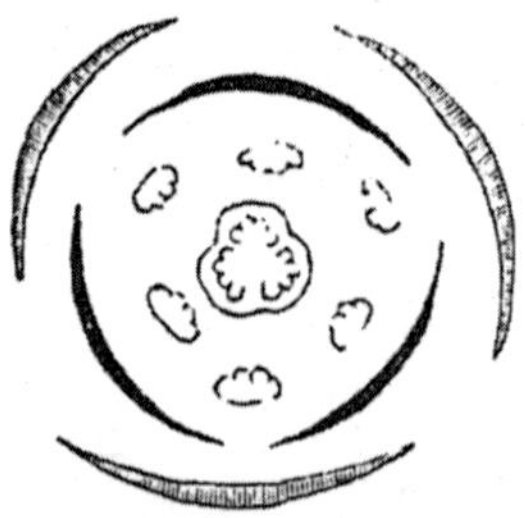

Fig. 288. Colchicum corm. *r.* Roots. *f.* Leaf. *a.* Shrivelled remains of last year's corm. *a''.* Corm of the present year. *a'''.* Commencement of the corm of next year.——*Fig.* 289. Flowering plant of the Colchicum or Meadow Saffron (*Colchicum autumnale*).——*Fig.* 290. Diagram of the flower of the same, with six divisions to the perianth, arranged in two whorls; six stamens; and a 3-celled ovary.

septicidal dehiscence (*fig.* 292). *Seeds* numerous; *embryo* minute, in fleshy albumen (*fig.* 294).

By Bentham and Hooker the plants of this order are now placed, according to the arrangement of Baker, in the Liliaceæ, and constitute the tribe *Colchiceæ* of that order. (See *Liliaceæ.*)

Diagnosis.—Herbs. Flowers regular, perfect or rarely unisexual. Perianth inferior, 6-partite or 6-leaved. Stamens 6; anthers extrorse. Ovary superior; style 3-partite. Fruit a septicidal capsule,

FIG. 291.

FIG. 292.

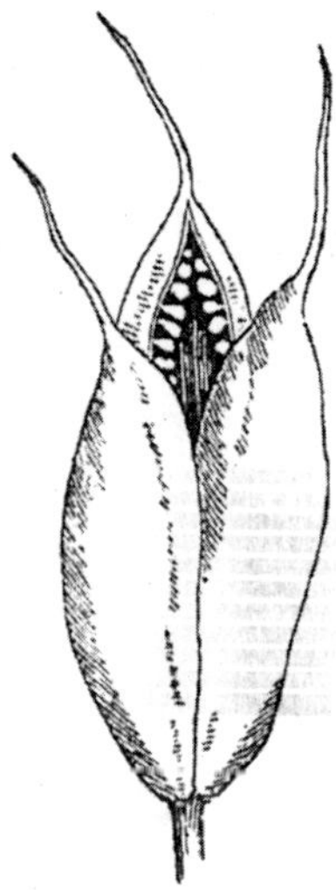

FIG. 293.

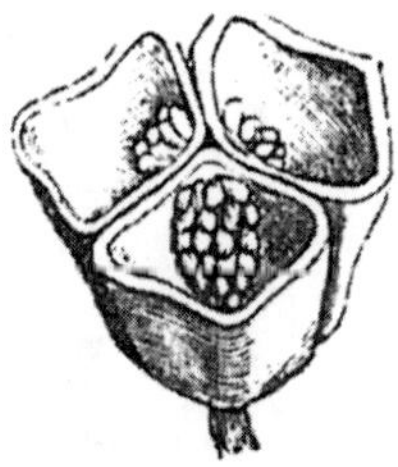

FIG. 294.

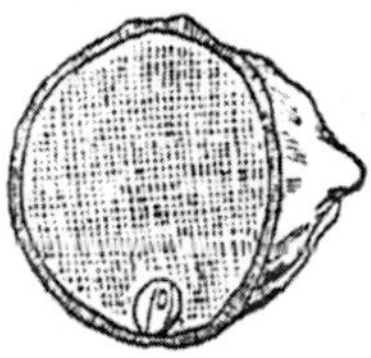

Fig. 291. The perianth cut open, showing the stamens of the Meadow Saffron (*Colchicum autumnale*) with their anthers turned towards the floral envelopes, and hence termed extrorse.——*Fig.* 292. Capsule of the same, showing septicidal dehiscence.——*Fig.* 293. Transverse section of the capsule.——*Fig.* 294. Vertical section of the seed.

3-celled, 3-valved, membranous. Seeds numerous, albuminous. *Illustrative Genera*:—Colchicum, *Linn.*; Tofieldia, *Hudson.*

3. Diclines.

AROIDACEÆ, the Arum Order.—Character.—*Herbs* or *shrubs* with commonly an acrid juice, and subterranean tubers, corms (*fig.* 295, *b*), or rhizomes. *Leaves* sheathing (*fig.* 295, *l*), usually net-veined, simple or rarely compound. *Flowers* unisexual (*fig.* 297) or bisexual, arranged on a spadix (*figs.* 296 and 297) within a spathe (*fig.* 296), or the spathe is absent. *Perianth* none (*fig.* 297), or composed of scales which are inferior. *Male*

FIG. 295.

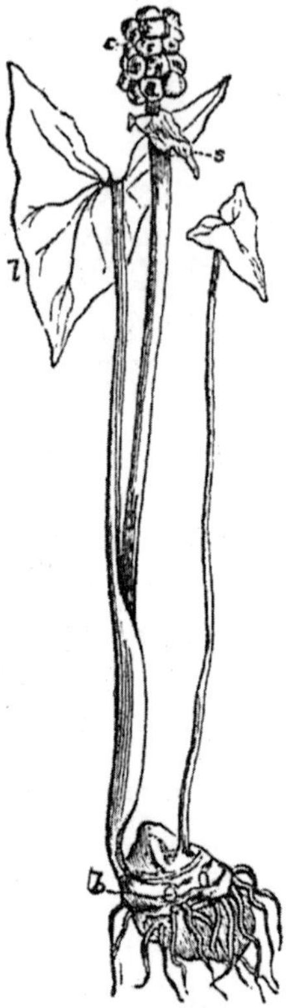

FIG. 296.

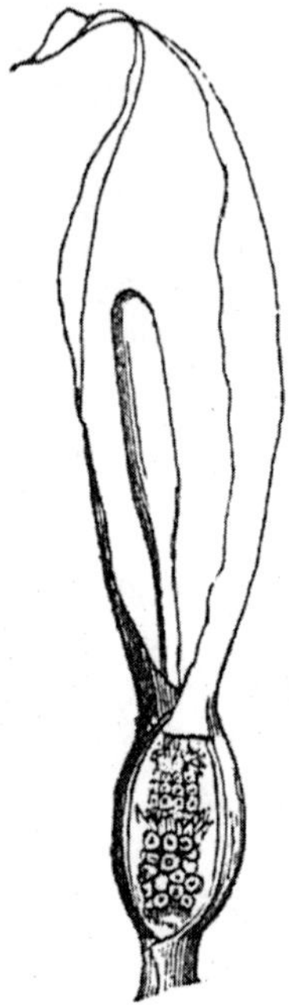

FIG. 297.

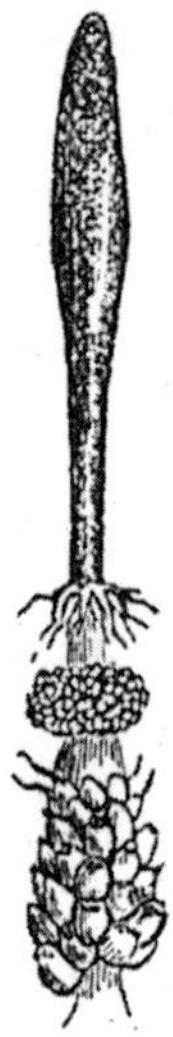

Fig. 295. A plant of the Cuckoo-pint (*Arum maculatum*) in fruit. *b.* Corm. *l.* Leaf. *s.* The remains of the spathe. *c.* Fruit.——*Fig.* 296. Spadix of Cuckoo-pint (*Arum maculatum*) enclosed in a spathe, a portion of which has been removed to show the flowers within it.——*Fig.* 297. The spadix of the same with the spathe removed; the flowers are all naked and unisexual, a number of pistillate flowers being below, above which are some rudimentary ovaries, then a number of sessile anthers, and above these are some staminodes or abortive stamens.

flower:—*Stamens* few or numerous; *anthers* extrorse, sessile (*fig.* 298), or upon very short filaments. *Female flower*:—*Ovary* (*fig.* 299) 1- or more celled. *Fruit* succulent (*fig.* 295, *c*). *Seeds* pulpy, with abundant mealy, fleshy, or horny albumen (*fig.* 300) or rarely exalbuminous; *embryo* various.

Diagnosis.—Flowers on a spadix, and with or without a true spathe. Flowers naked, unisexual and monœcious, or perfect, and then frequently with a scaly inferior perianth. Anthers extrorse. Fruit succulent.

Division of the Order and Illustrative Genera:—The order may be divided into two sub-orders as follows:—

Sub-order 1. *Aroideæ* or *Araceæ.*—Flowers unisexual, monœcious. Spadix surrounded by a spathe. Perianth none. *Illustrative Genus:*—Arum, *Linn.*

FIG. 298.

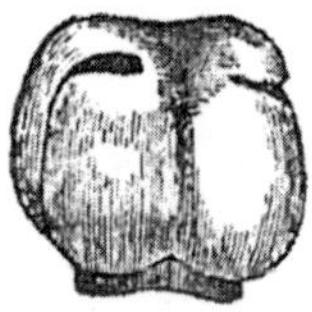

FIG. 299.

FIG. 300.

Fig. 298. Stamen of the Cuckoo-pint (*Arum maculatum*), consisting simply of an anther which is sessile upon the thalamus.——*Fig.* 299. Vertical section of the pistil of the same.——*Fig.* 300. Vertical section of the seed.

Sub-order 2. *Acoreæ* or *Orontieæ.*—Flowers perfect. Spadix surrounded by a spathe or naked. Perianth absent, or more generally present, and then scaly. *Illustrative Genus:*—Acorus, *Linn.*

This order was divided by Lindley into two separate orders—the Araceæ and Acoraceæ, on account of the perfect flowers of the latter. In accordance, however, with the more general views of botanists, we make but one order, and place the two orders of Lindley as sub-orders. It should be noticed, however, that in adopting this arrangement, the sub-order *Araceæ* is alone illustrative of the Diclines in our division of the Monocotyledones.

Sub-class II. *Glumaceæ* or *Glumiferæ.*

CYPERACEÆ, the Sedge Order.—Character.—Grass-like (*fig.* 301) or Rush-like usually perennial *herbs.* *Stems* solid, without joints or partitions, frequently angular (*fig.* 302). *Leaves* without ligules, and with entire or undivided sheaths round the stem (*fig.* 302). *Flowers* spiked, imbricate, perfect (*fig.* 303) or unisexual (*figs.* 304 and 305), each arising from the axil

FIG. 301.

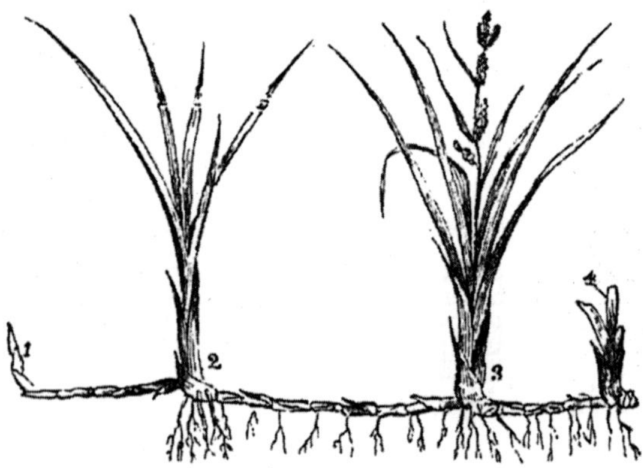

FIG. 302.

Fig. 301. Creeping stem of the Sand Carex (*Carex arenaria*). 1. Terminal bud by which the stem continues to elongate. 2, 3, 4. Shoots produced from axillary buds.——*Fig.* 302. A portion of the angular stem of a species of *Carex*, with a closed sheath.

of 1 (*figs.* 304, *g*, and 305)—3 bracts or glumes. (*The lowermost glumes are frequently empty, that is, without flowers in their axils.*) *Perianth* absent, or existing in the female flowers in the form of a tube (*perigynium*) (*fig.* 305, *u*), or as hypogynous scales or bristles (*fig.* 303, *b*). *Stamens* hypogynous (*fig.* 303), 1—12, commonly 3 (*figs.* 303 and 304); *anthers* 2-celled, innate (*figs.* 303 and 304). *Ovary* 1-celled, superior (*fig.* 303, *o*), with 1 erect anatropous ovule. *Fruit* indehiscent, 1-seeded (*fig.* 306). *Seed* with fleshy or mealy albumen (*fig.* 306, *alb*); *embryo* lenticular (*figs.* 306, *pl*, and 307), enclosed in the base of the albumen.

FIG. 303.

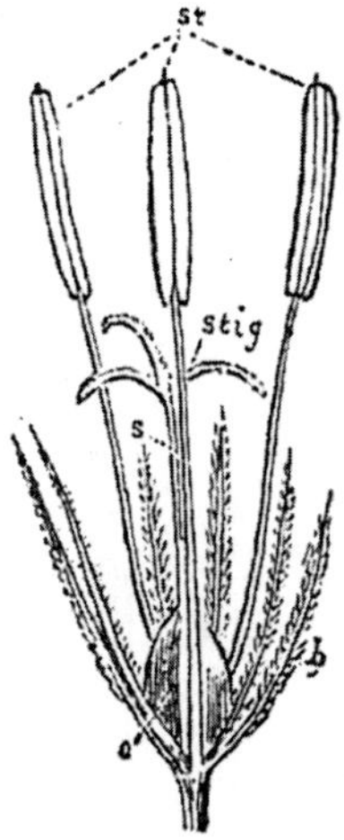

FIG. 304.

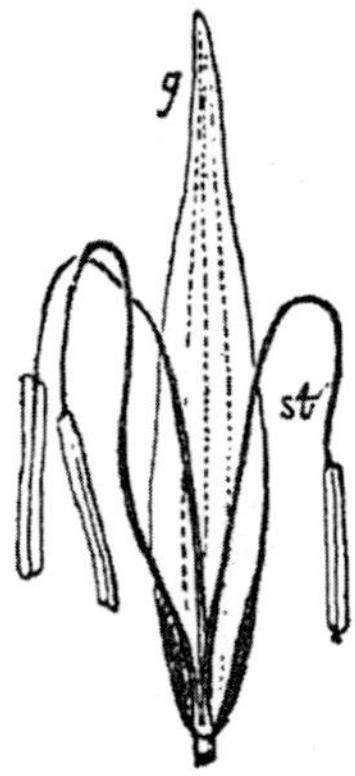

FIG. 305.

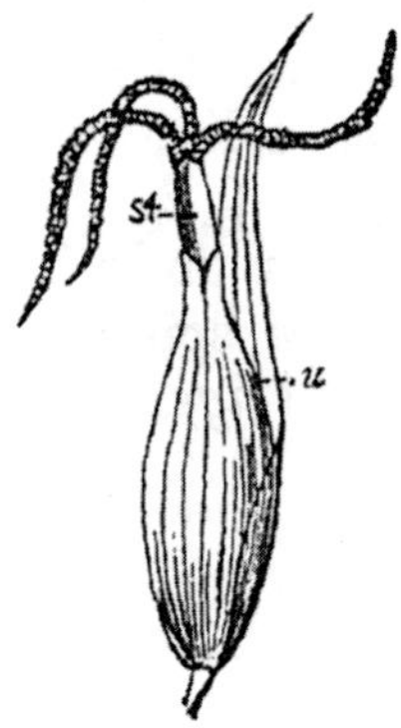

Fig. 303. Perfect or bisexual flower of a species of Club-rush (*Scirpus*), the glume having been removed. *b.* Hypogynous setæ or bristles forming a kind of perianth. *st.* Hypogynous stamens with 2-celled innate anthers. *o.* Ovary. *s.* Style. *stig.* Stigmas. ——*Fig.* 304. Staminate flower of a species of *Carex*. *st.* Stamens, with long filaments and pendulous innate anthers. *g.* Glume. ——*Fig.* 305. Pistillate flower of a species of *Carex*, consisting of a glume at the base, and a pistil surrounded by an urn-shaped tube (*perigynium*), *u.* *st.* Style, terminated by three stigmas.

FIG. 306.

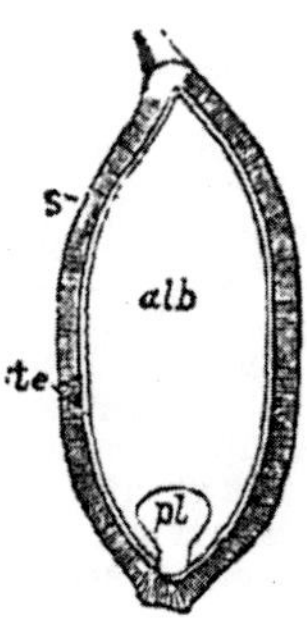

Fig. 306. Vertical section of the fruit of a species of *Carex*. *s.* Pericarp. *te.* Integuments of the seed. *alb.* Albumen. *pl.* Embryo.——*Fig.* 307. Embryo of a species of *Carex* removed from the albumen. *a.* Lateral swelling. *r.* Radicle. *c.* Cotyledon. *f.* Slit corresponding to the plumule.

FIG. 307.

Diagnosis.—Grass-like or Rush-like herbs with solid and usually angular stems. Leaves without ligules and with closed sheaths. Stamens few, hypogynous;

anthers innate, 2-celled. Ovary superior, 1-celled; ovule solitary, anatropous. Fruit indehiscent, 1-celled, 1-seeded. Embryo enclosed in the base of albumen. *Illustrative Genera*:—Carex, *Linn.*; Scirpus, *Linn.*; Cyperus, *Linn.*

GRAMINACEÆ, the Grass Order.—Character.—*Herbs, shrubs*, or *arborescent plants*, with round, com-

FIG. 308.

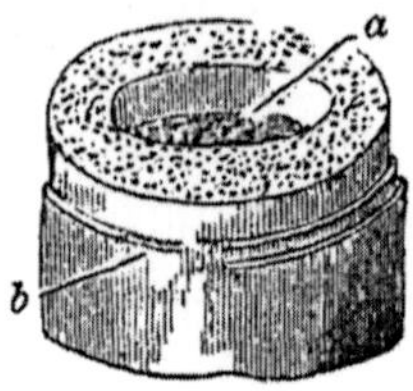

FIG. 309.

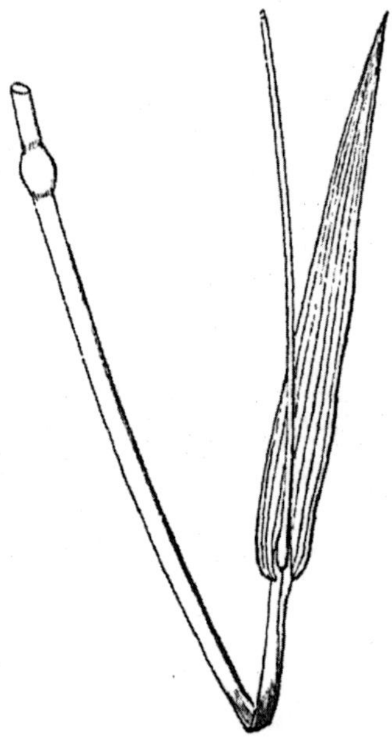

IG. 310.

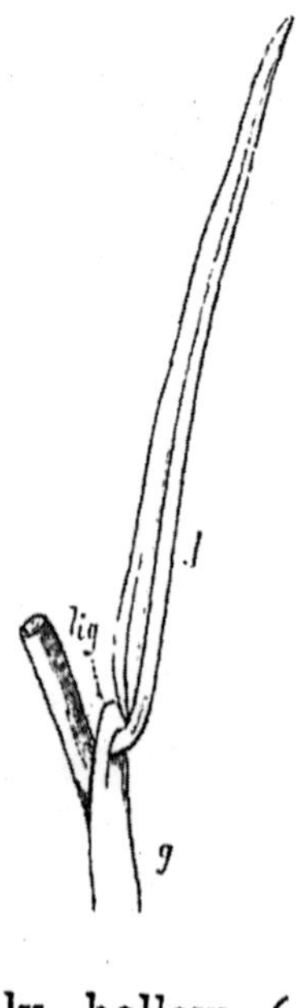

Fig. 308. Transverse section of the stem of the common Reed (*Phragmites communis*). *a.* Cavity closed at the bottom by a partition. *b.* Annular scar indicating the point (*node*) where the leaf was attached.——*Fig.* 309. A portion of the stem of the Cat's-tail Grass (*Phleum pratense*), bearing a leaf with parallel veins, and a split sheath.——*Fig.* 310. Stem of a Grass with a leaf attached. *l.* Blade. *g.* Split sheath. *lig.* Ligule.

monly hollow (*fig.* 308, *a*), jointed stems. *Leaves* alternate, with parallel veins (*fig.* 309), and split sheaths (*figs.* 310, *g*, and 309), and with a ligule at the base of the lamina (*fig.* 310, *lig*). *Flowers* perfect or unisexual, arranged in spiked (*fig.* 311), panicled (*fig.* 312), or racemose locustæ; or solitary. *No true perianth*, its place being supplied by imbricated bracts, of which there are commonly 2, called *glumes*, or rarely 1; these glumes are placed at the base of the solitary flower, or at

FIG. 311.

FIG. 312.

Fig. 311. Inflorescence of Wheat (*Triticum sativum* or *vulgare*), consisting of numerous sessile spikelets arranged on a common axis (*rachis*).——*Fig.* 312. Branched or panicled inflorescence of the Oat (*Avena sativa*).

the base of each locusta (*figs.* 313, *gl*, *gl*, and 314, *gl*, *gl*, and 315, *ge*, *gi*). Occasionally the glumes are absent. Each flower is also usually furnished with two other alternate bracts (*paleæ*) (*figs.* 315, *pe*, *pi*, and 316 and 317) (or sometimes the inner palea, *pi*, is wanting), the outer palea is frequently known as the *flowering glume*; and 2 or 3 hypogynous scales (*lodiculæ*, *squamulæ*, or *glumellules*) (*figs.* 314, *p*, *p*, 318, *p*, and 320, *sp*), representing a perianth; these scales also are occasionally absent. *Stamens* hypogynous, 1—6, usually 3 (*figs.* 316–318); *filaments* capillary (*figs.* 316 and 319); *anthers* versatile (*figs.* 316, 319, and 320). *Ovary* superior (*figs.* 318, *o*, and 320), 1-celled, with a solitary ascending ovule; *stigmas* feathery (*figs.* 318, *s*, *s*, and 320) or hairy. *Fruit* a caryopsis (*figs.*

FIG. 313.

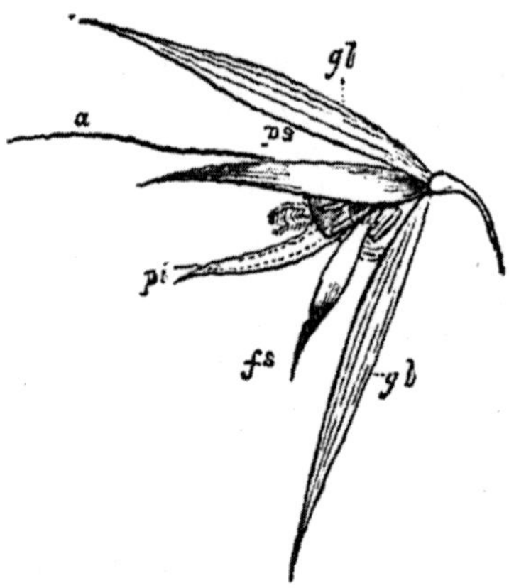

Fig. 313. Locusta or spikelet of the Oat (*Avena sativa*). *gl. gl.* Glumes. *ps, pi. Paleæ* or *Pales*. *a*. Awn arising from the dorsum of the outer pale, or flowering glume, *ps*. *fs*. An abortive flower.

321 and 322). *Seed* with mealy albumen (*fig.* 322, *a*); *embryo* lenticular (*fig.* 323), lying on one side at the base of the albumen (*fig.* 322, *c*, *g*, *r*).

Diagnosis.—Leaves alternate, with split sheaths, and a ligule at the base of the lamina. Flowers generally arranged in spikelets or locustæ, or rarely solitary. Flowers glumaceous; paleæ usually two in each flower. Stamens hypogynous, few, usually 3, with capillary filaments, and versatile anthers. Ovary superior; stigmas feathery or hairy. Fruit a caryopsis. Seed with mealy albumen, with the embryo on one side at the base. *Illustrative Genera*:—Panicum, *Linn.*; Anthoxan-

FIG. 314.

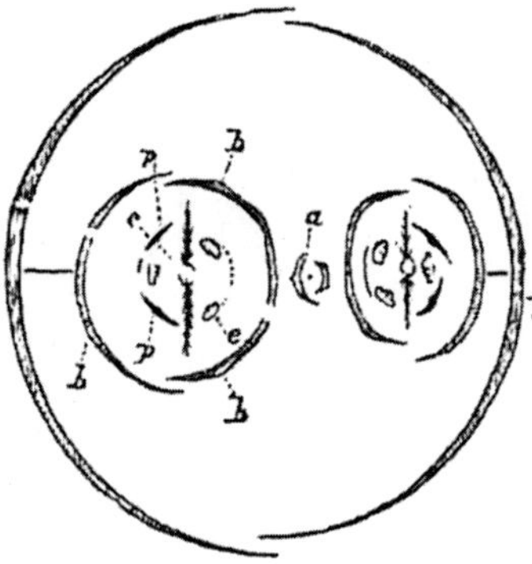

FIG. 315.

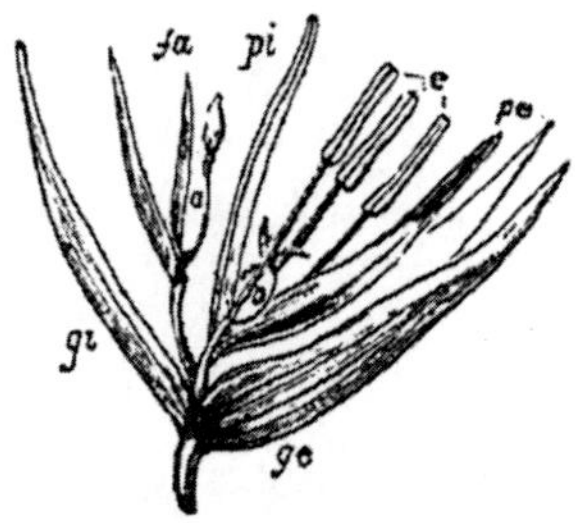

FIG. 316.

FIG. 317.

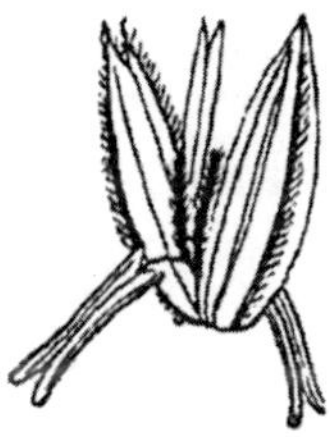

FIG. 318.

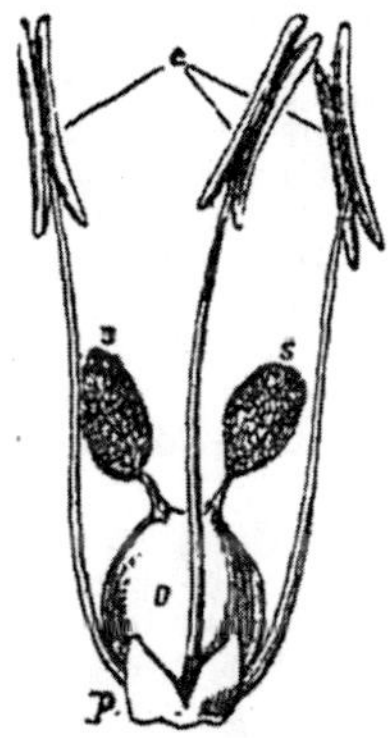

Fig. 314. Diagram of a spikelet of the Oat (*Avena sativa*). (From Le Maout.) *gl*, *gl*. Two glumes, enclosing two perfect flowers, and one, *a*, abortive. *b*. The outer palea or flowering glume. *b*, *b*. The inner palea, which seems to be formed of two united. *p*, *p*. Two scales (*squamulæ* or *glumellules*); the dotted curved line on the right marks the position of a third abortive scale. *e*. Stamens. *c*. Ovary.——*Fig.* 315. A spikelet (locusta) of the Oat (*Avena sativa*). *ge*. Outer glume. *gi*. Inner glume. *pe*. Outer palea or flowering glume of the fertile flower. *pi*. Inner palea of the same. *e*. Stamens. *o*. Ovary. *fa*, and *a*. Abortive flowers.——*Fig.* 316. One of the florets of a species of Meadow Grass (*Poa pratensis*).——*Fig.* 317. One of the florets of the Hard Fescue Grass (*Festuca duriuscula*).——*Fig.* 318. Fertile flower of the Oat, without the paleæ. *p*. Glumellules. *e*. Stamens. *o*. Ovary. *s*, *s*. Feathery stigmas.

thum, *Linn.*; Phleum, *Linn.*; Agrostis, *Linn.*; Dactylis, *Linn.*; Bromus, *Linn.*

FIG. 319.

FIG. 320.

Fig. 319. A locusta of Wheat (*Triticum sativum*) consisting of severa flowers, the stamens of which have very long capillary filaments, and versatile pendulous anthers. The anthers are bifurcated at each extremity, and resemble somewhat the letter *x* in form. ——*Fig.* 320. Pistil of Wheat surrounded by three stamens, and two lodiculæ or squamulæ, *sp*. Two feathery styles or stigmas arise from the top of the ovary.

FIG. 321.

FIG. 322.

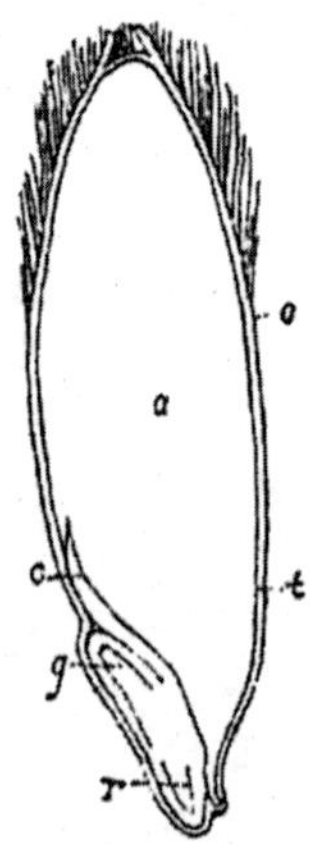

FIG. 323

Fig. 321. Caryopsis or fruit of the Oat.——*Fig.* 322. The same, cut vertically. *o*. Pericarp. *t*. Integuments of the seed. *a*. Albumen. *c*. Cotyledon. *g*. Plumule. *r*. Radicle.——*Fig.* 323. The embryo of the Oat. *a*. Lateral swelling. *c*. Cotyledon. *r*. Radicle. *f*. Slit corresponding to the plumule.

SUB-KINGDOM II.

CRYPTOGAMIA, ACOTYLEDONES, OR FLOWERLESS PLANTS.

CLASS III. ACOTYLEDONES.

Sub-class I. *Acrogenæ* or *Cormophyta.*

FILICES, the Fern Order.—Character.—*Herbs* with *rhizomatous stems* (*fig.* 324); or *arborescent plants* with cylindrical stems, usually unbranched (*fig.* 325), but sometimes forked (*fig.* 326). *Leaves*, or *fronds* as they are commonly called, arising irregularly from the rhizome (*fig.* 324) or placed in tufts at the apex of the cylindrical stem or caudex (*fig.* 325); almost always circinate in vernation (*figs.* 324 and 325); simple (*fig.* 327, *a*) or compound (*figs.* 324 and 325). *Fructification* consisting of *sporangia* or *capsules* (*figs.* 328 and 329), collected in heaps (*sori*), which are placed usually on the under surface (*figs.* 328, *sp*, and 329, *s*, *s*) or at the margins of the fronds (*fig.* 327, *b*), or rarely on the upper surface, or occasionally arranged in a spiked manner on a simple or branched rachis (*fig.* 330); the *sori* are either naked (*fig.* 328) or covered by a membranous scale (*indusium*) (*fig.* 329). *Sporangia* stalked (*fig.* 331, *s*) or sessile (*fig.* 327, *b*), and either annulate (*fig.* 331) or exannulate (*fig.* 327, *b*). *Spores* enclosed in the sporangia (*fig.* 331, *s*).

Division of the Order and Illustrative Genera.—This order has been variously divided by botanists. The more common arrangement is into three sub-orders called Polypodieæ, Danæeæ, and Ophioglosseæ. Their characters are as follows:—

Sub-order 1. *Polypodieæ* or *Polypodiaceæ.*—The Polypody Sub-order or Ferns proper.—Fronds circinate in vernation (*figs.* 324 and 325). Sporangia more or less annulate (*fig.* 331, *s*), usually collected in

FIG. 324.

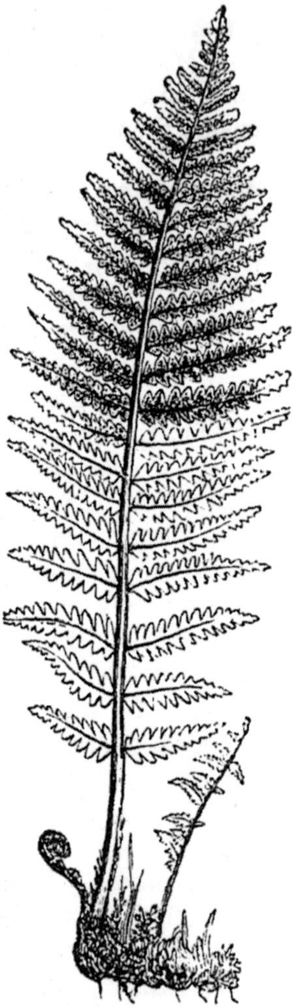

FIG. 325.

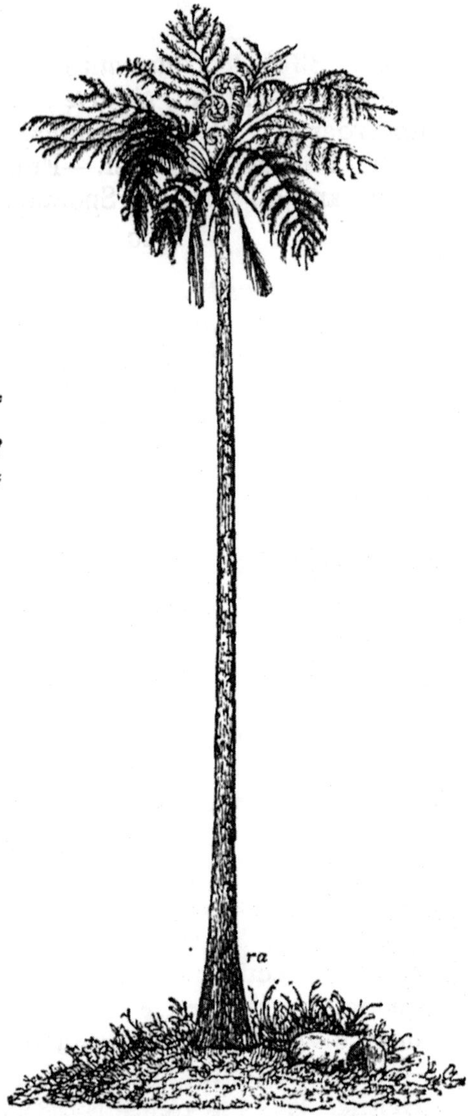

Fig. 324. The Male-Fern (*Aspidium Filix-mas*).——*Fig.* 325. A Tree-fern, showing a tuft of leaves (*fronds*) at the apex of a cylindrical stem, which is enlarged at its base, *ra*, by the development of a mass of adventitious roots.

sori on the under surface (*figs.* 328 and 329), or at the margins of the fronds, or occasionally arranged in a spiked manner on a simple or branched rachis (*fig.* 330). *Illustrative Genera:* — Polypodium, *Linn.*; Aspidium, *Swartz*; Osmunda, *Linn.*

Sub-order 2. *Danæeæ*, *Danæaceæ*, or *Marattiaceæ*.—The Danæa Sub-order.—Fronds circinate in vernation, and all fertile. Sporangia arising from, or im-

FIG. 326.

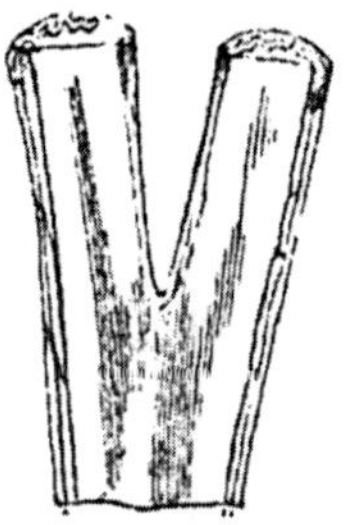

FIG. 327.

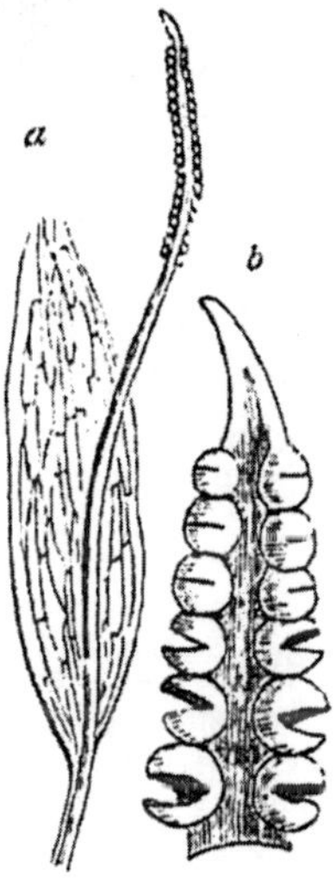

FIG. 328.

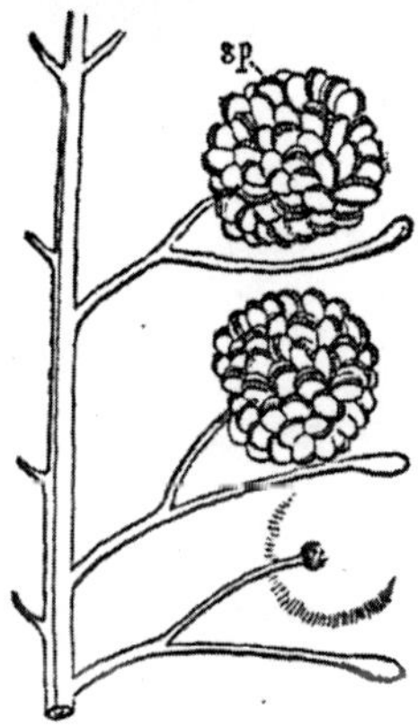

Fig. 326. Vertical section of the forked stem of a Tree-fern.——*Fig.* 327. *a.* Barren and fertile fronds of the Common Adder's-tongue (*Ophioglossum vulgatum*). *b.* Portion of the fertile frond of the same, with 2-valved, distinct, burst sporangia or capsules on its margins.——*Fig.* 328. A portion of the frond of the common Polypody (*Polypodium vulgare*), showing two sori springing from its veins on the under surface. The sori are naked, and consist of a number of sporangia or capsules, *sp*, in which the spores are contained.

bedded in, the under surface or back of the fronds, more or less united, exannulate. *Illustrative Genera:* —Danæa, *Smith*; Marattia, *Smith*. There are no British plants in this sub-order.

Sub-order 3. *Ophioglosseæ* or *Ophioglossaceæ*.—The Adder's-tongue Sub-order.—Fronds not circinate in vernation (*fig.* 327), barren or fertile (*fig.* 327, *a*, *b*).

Sporangia arranged in a spike-like form on the margins of a contracted frond (*fig.* 327, *b*), distinct, 2-valved, exannulate. *Illustrative Genera:*—Ophioglossum, *Linn.*; Botrychium, *Swartz.*

FIG 329.

FIG. 330.

FIG. 331.

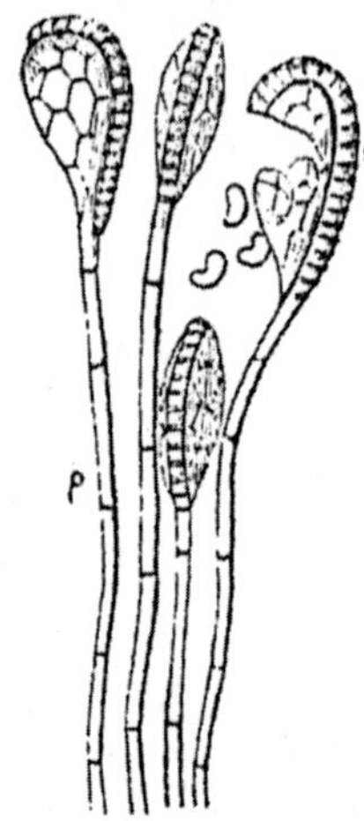

Fig. 329. Portion of the under surface of a frond of the Male-fern (*Aspidium Filix-mas*), with two sori, *s*, *s*, covered by an indusium of involucre.——*Fig.* 330. Portion of the frond of the Royal or Flowering-fern (*Osmunda regalis*), with its sporangia or capsules, arranged in a spiked manner on a branched rachis.——*Fig.* 331. Sporangia or capsules of a Fern. *s*. Sporangium supported on a stalk, *p*, and surrounded by a ring or annulus, which is a continuation of the stalk. One capsule is represented as burst on its side, and the contained spores in the act of being scattered.

MUSCI, the Moss Order.—Character.—*Cellular plants* (*figs.* 332–334), terrestrial or aquatic, with erect or creeping stems, and usually spirally imbricated leaves (*figs.* 332–335). *Reproductive organs* of two kinds, called *antheridia* and *archegonia*, which are either placed on the same or on separate plants (*figs.* 332 and 333); hence these plants are *monœcious* or *diœcious*. The *antheridium* (*fig.* 336) is a more or less rounded, elliptic, or cylindrical sac, containing, when mature, a number of minute cells, *c* (*zoothecæ*), each of which encloses a spirally twisted filament (*antherozoid*). The *archegonium* is usually a flask-shaped body (*fig.*

337), which after fertilisation develops an urn-shaped sporangium (*figs.* 338–341), with a central columella (*fig.* 343); the space between which and the walls of

FIG. 332.

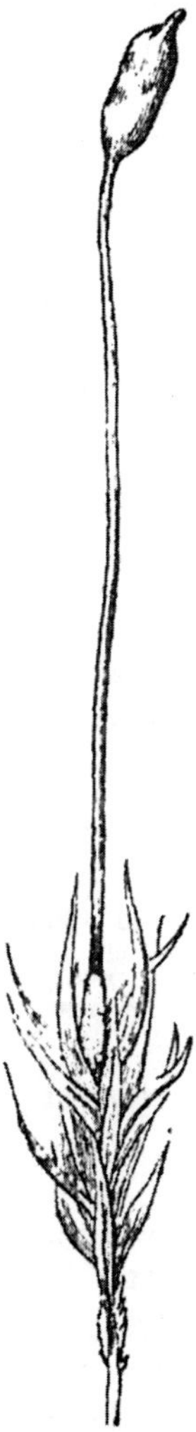

FIG. 333.

FIG. 334.

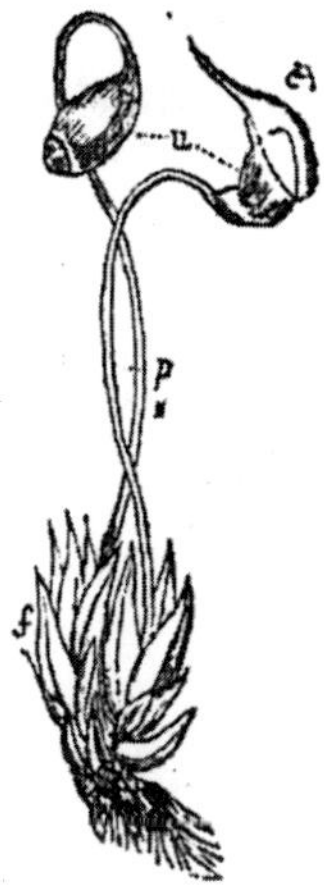

Fig. 332. Female plant of the Hair-moss (*Polytrichum commune*), with its leaves, stem, and fructification.——*Fig.* 333. The male plant of the same, with its stem and leaves, and terminated by the male organs (*antheridia*).——*Fig.* 334. The Hygrometric Cord-moss (*Funaria hygrometrica*). *f.* Perichætial leaves. *p.* Stalks or *setæ*, each of which supports a sporangium, *u*, covered by a calyptra,

the sporangium being occupied by spores, without any elaters among them. The *sporangium* or *capsule* is commonly placed on a stalk (*seta*) (*figs.* 338, *t*, and

334, *p*), or occasionally it is sessile (*fig*. 335), and at first is covered by a hood (*calyptra*) (*figs*. 334, *c*, and 339, *c*), beneath which is a kind of lid (*operculum*) (*figs*. 340, *o*, and 341). The sporangium usually opens when ripe in a transverse manner from the separation of the operculum (*figs*. 340, *o*, and 341), or sometimes by splitting vertically into four equal valves, which are connected at the summit by the persistent operculum

FIG. 335.

FIG. 336.

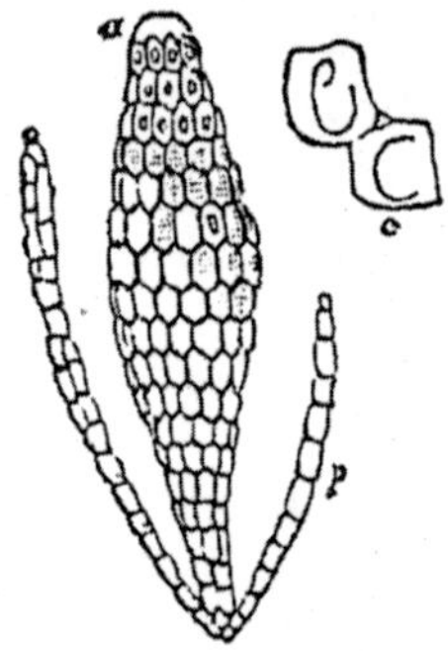

FIG. 337.

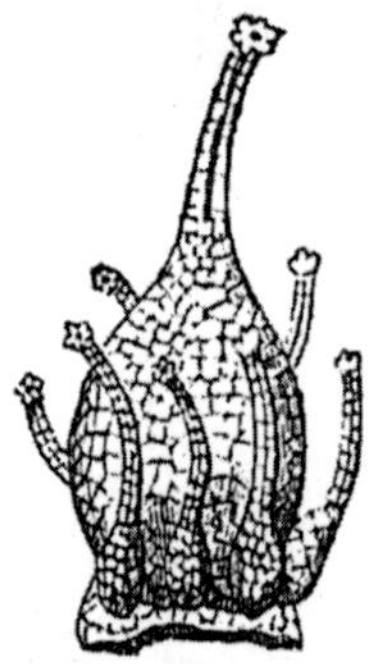

Fig. 335. A portion of *Andræa rupestris*, much magnified. The stem is erect, with numerous small imbricated leaves, and a terminal sporangium, which is destitute of a seta. *a*. Sporangium after dehiscence, showing the four equal valves of which it is composed connected at the summit by the persistent operculum. The valves are seen to have dehisced vertically.——*Fig*. 336. Antheridium, *a*, of the Hair-moss (*Polytrichum commune*), containing a number of cells, *c*, in each of which there is a single *antherozoid*. *p*. *Paraphyses*, surrounding the antheridium.——*Fig*. 337. Archegonium or pistillidium of a moss surrounded by *paraphyses*.

(*fig*. 335, *a*); or rarely it dehisces irregularly. At the dehiscence of the sporangium, its mouth (*stoma*) is seen to be either surrounded by a peristome, consisting of one (*aploperistomous*) or two rows (*diploperistomous*) of teeth (*fig*. 340, *p*); or the mouth is naked (*gymnostomous*) (*fig*. 341).

Division of the Order and Illustrative Genera.—This order is commonly divided into four sub-orders, the principal distinctive characters of which are as follows:—

Sub-order 1. *Sphagnaceæ* or *Sphagneæ.*—Bog-mosses.

FIG. 339.

FIG. 340.

FIG. 338.

FIG. 341.

FIG. 342.

FIG. 343.

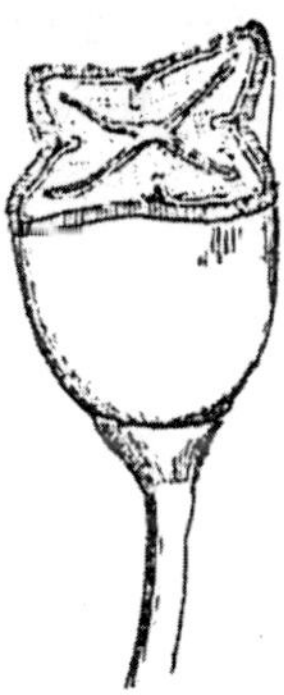

Fig. 338. *Coscinodon pulvinatus. sp.* Sporangium or capsule enclosed in the *calyptra. t. Seta* or stalk. *v.* Vaginule. From Henfrey.——*Fig.* 339. Sporangium of the Extinguisher-moss (*Encalypta vulgaris*) before dehiscence. *u.* Sporangium covered by a transparent calyptra, *c*, and supported on a seta. *s.* Beneath the calyptra is seen the lid or operculum, *o.*——*Fig.* 340. The sporangium, *u*, of 339, after dehiscence. The calyptra and operculum, *o*, being removed, the *peristome*, *p*, may be seen. ——*Fig.* 341. *Pottia truncata*, showing the separation of the operculum from the sporangium. From Henfrey.——*Fig.* 342. Sporangium, *u*, of the Hair-moss (*Polytrichum commune*), deprived of its calyptra and operculum. *p.* Peristome. *e. Epiphragma* or *tympanum.*——*Fig.* 343. Transverse section of a sporangium of the same, showing the central columella surrounded by free spores.

—Sporangium globular, surrounded at the base by the calyptra; the columella does not reach to the apex of the sporangium. *Illustrative Genus:*—Sphagnum, *Dill.*, is the only genus.

Sub-order 2. *Andræaceæ* or *Andræeæ*.—Split-mosses.—Sporangium splitting vertically into four valves, but remaining connected at the summit (*fig.* 335, *a*). *Illustrative Genus:*—Andræa, *Ehr.*

Sub-order 3. *Phascaceæ* or *Phasceæ*.—The sporangium does not burst; the spores escaping by the decay of the wall of the capsule. *Illustrative Genus:*—Phascum, *Linn.*

Sub-order 4. *Bryaceæ*.—Urn-mosses.—Sporangium, which is generally borne upon a seta of considerable length, dehiscing transversely by the separation of the operculum (*figs.* 334, 340, and 341). *Illustrative Genera:*—Funaria, *Hedgw.*; Polytrichum, *Linn.*

Sub-class II. *Thallophyta* or *Thallogenæ*.

FUNGI, the Mushroom Order.—*Diagnosis.* Plants

FIG. 344.

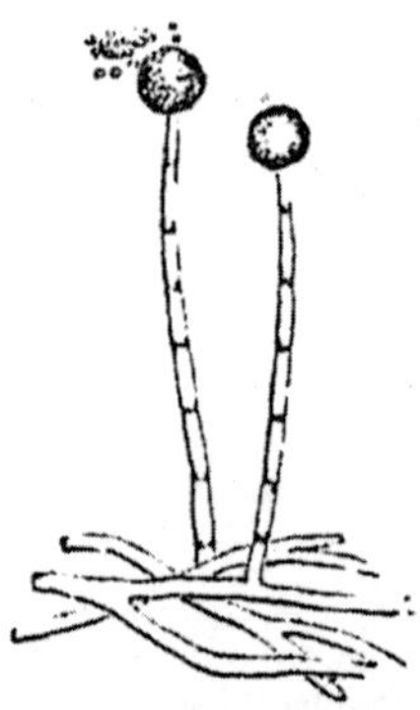

FIG. 345.

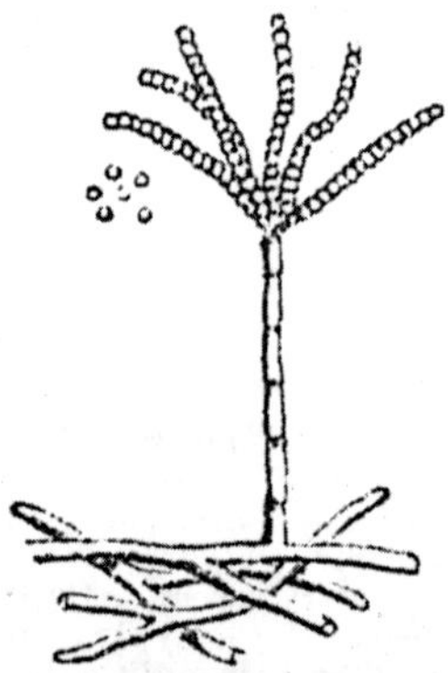

Fig. 344. A species of Mould (*Mucor*), with branched mycelium (*hyphal tissue*) below, from which two stalks are seen to arise, each of which is terminated by a sac (*sporangium* or *ascus*), from which a number of minute bodies (*spores*) are escaping.—*Fig.* 345. Another Mould (*Penicillium glaucum*), with branched mycelium (*hyphal tissue*), and stalk bearing several rows of cells, which are the germinating spores (*conidia*).

formed of hyphal tissue (*figs.* 344 and 345), producing their fructification in the air; growing in or upon decaying organic matters (in which case they are called *saprophytes*), or on living organisms (when they are termed *parasites*), and nourished through their vegetative structure called the *spawn* or *mycelium* (*fig.* 344, 345, and 346, *my*). The Fungi, as here defined, are also destitute of green colouring matter and starch. *Fructification* various. (See the author's 'Student's Guide to Structural, Morphological, and Physiological Botany,' pp. 350–361.)

Division of the Order and Illustrative Genera:—For

FIG. 346.

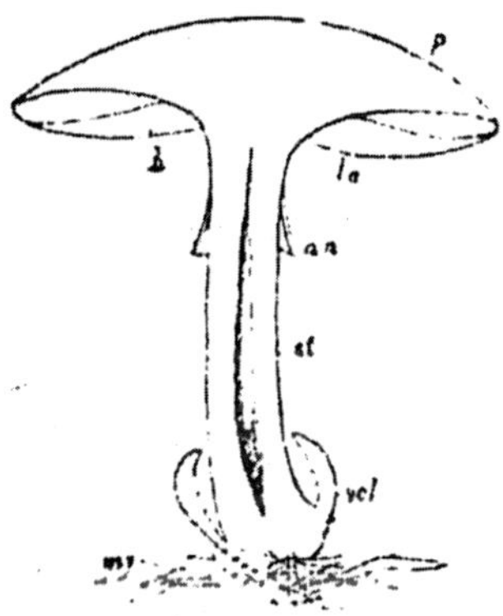

Fig. 346. Vertical section of the common Mushroom (*Agaricus campestris*). *my*. Mycelium. *vol*. Remains of volva. *st*. Stipe. *an*. Annulus. *h*. Hymenium with its gills or lamellæ, *la*. *p*. The pileus.

a notice of the groups into which the order has been divided, and their illustrative genera, reference may also be made to the author's 'Student's Guide,' as above.

LICHENES, the Lichen Order.—Character.—*Perennial plants*, composed of hyphal tissue (*figs.* 347 and 348) resembling that of Fungi, but its constituent cells are dry and firm, and enclose the cells, known as gonidia, *gon*, which contain chlorophyll, and are now very commonly regarded as minute Algæ, upon which the Fungus is parasitic. The whole is arranged so as to form a foliaceous (*fig.* 350), somewhat woody, scaly, crustaceous, or leprous, thallus;

living and fructifying in the air, and growing on the bark of trees, or on old palings, walls, &c., or on stones, or on the exposed surface of rocks; usually epiphytic, but sometimes parasitic, and commonly presenting a dry, shrivelled, more or less lifeless appearance. *Reproduction* either vegetative by means of

FIG. 347.

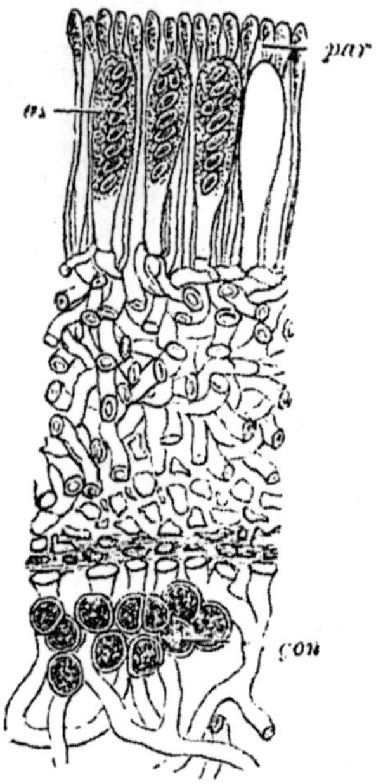

FIG. 348.

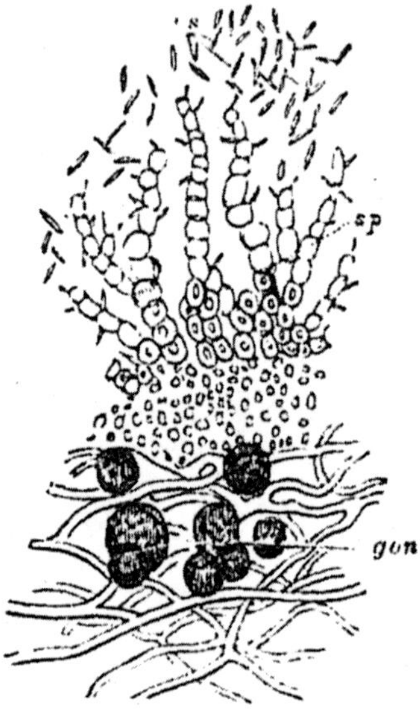

Fig. 347. Section of the thallus of *Cetraria islandica* through an apothecium. *as.* Asci, three of which contain ascospores. *par.* Paraphyses. *gon.* Gonidia enclosed in hyphal tissue. After Berg and Schmidt.——*Fig.* 348. Highly magnified fragment from the wall of a spermogonium of *Parmelia parietina*. *sp.* Articulated sterigmata or spermatophores. *s.* Spermatia. *gon.* Gonidia enclosed in hyphal tissue. After Henfrey.

soredia; or by true *fructification*, consisting of, 1. *apothecia*, which are sessile or stalked, and generally of a rounded (*fig.* 350, *ap*) or linear form (*fig.* 349), and composed of *asci* or *thecæ* (*fig.* 347, *as*), enclosing 4, 8, or 16 spores; 2. of *spermogonia* (*figs.* 350, *sp*, and 351, *sp*), containing *spermatia* (*figs.* 348, *s*, and 351, *s*); and

3. of, very rarely, *pycnidia* enclosing *stylospores*. (For a detailed account of the fructification of Lichens, see the author's 'Guide to Structural, Morphological, and Physiological Botany,' pages 361–365). *Illustrative Genera:*—

FIG. 349.

FIG. 350.

FIG. 351.

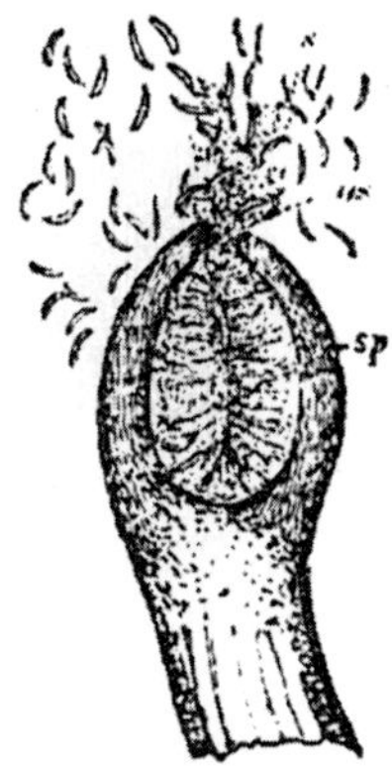

Fig. 349. Thallus of *Opegrapha atra*, showing linear apothecia, which are termed lirellæ.——*Fig.* 350. Portion of the thallus of *Parmelia parietina*, with young round *apothecia*, *ap*, and *spermogonia*, *sp*. After Henfrey.——*Fig.* 351. Vertical section of a spermogonium of *Cladonia rangiferina*. *sp*. Spermatophores. *os*. Ostiole or pore, from which the spermatia, *s*, are escaping.

Opegrapha, *Pers.*; Verrucaria, *Pers.*; Lecidea, *Ach.*; Cladonia, *Hoffm.*; Peltigera, *Hoffm.*; Usnea, *Hoffm.*

ALGÆ, the Seaweed Order.—*Diagnosis.*—*Parenchymatous plants*, growing in salt or fresh water, or in moist situations. The thallus is foliaceous and branched (*fig.* 352), or filamentous (*figs.* 353–355), or pulverulent. Many Algæ are microscopic, and others are of large size. In colour they are usually greenish, rose-coloured, or brown. They are reproduced in various ways. (See the author's 'Student's Guide to Structural, Morphological, and Physiological Botany,' pages 368–375.)

Division of the Order and Illustrative Genera:—This order is commonly divided into three sub-orders, which are frequently regarded as distinct natural orders: these are known under the names of the *Melanosporeæ*, *Melanospermeæ*, or *Fucoideæ*; *Rhodosporeæ*, *Rhodospermeæ*, or *Florideæ*; and *Chlorosporeæ*, *Chlorospermeæ*, or *Confervoideæ*. To these sub-orders may be added two others, called respectively *Diatomaceæ* and *Volvocineæ*. Numerous other arrangements of the Algæ have been proposed of late years, but as these must be regarded as transitional, we have in this little work retained the above-named sub-orders from their being generally used in this country in works treating practically of this order of plants. Their distinctive characters may be briefly described as follows:—

FIG. 352.

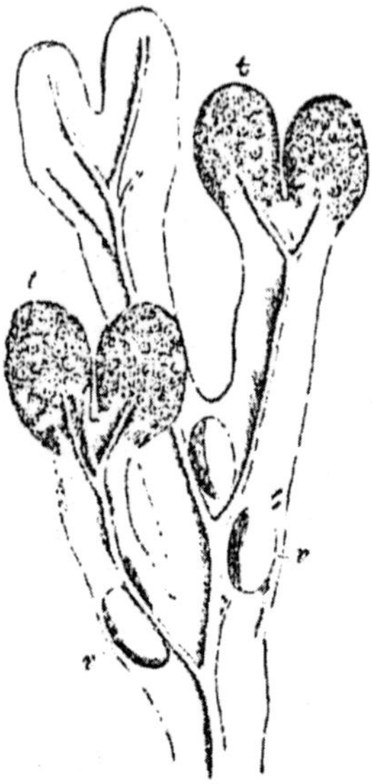

Fig. 352. Dichotomously-branched thallus or thallome of the common Bladder Sea-weed (*Fucus vesiculosus*). *t, t.* The fructification. *v, v.* Bladders of air.

Sub-order 1. *Melanosporeæ*, *Melanospermeæ*, *Fucoideæ*. or *Brown-coloured Algæ.*—Multicellular Algæ, growing in salt water, forming a foliaceous (*fig.* 352) or filamentous thallus, and of an olive-green or olive-brown colour. *Illustrative Genera*:—Sargassum, *Rumph.*; Fucus, *Linn.*

Sub-order 2. *Rhodosporeæ*, *Rhodospermeæ*, *Florideæ*, or *Rose-coloured Algæ.*—Marine multicellular Algæ, with a foliaceous or branched filamentous thallus, and of a reddish-purple, rose-coloured, or reddish-brown colour. *Illustrative Genera*:—Corallina, *Tourn.*; Chondrus, *Grev.*; Porphyra, *Agh.*

Sub-order 3. *Chlorosporeæ*, *Chlorospermeæ*, *Confervoideæ*, or *Green-coloured Algæ.*—Unicellular or multicellular Algæ (*fig.* 354), growing in fresh or salt water, or in moist situations; usually of a bright green colour, or rarely red. *Illustrative Genera*:—Conferva, *Plin.*; Palmella, *Agh.*; Spirogyra, *Link.*

Sub-order 4. *Diatomaceæ.*—Brittleworts.—The following diagnosis is from Henfrey:—' Microscopic unicellular plants, occurring isolated or in groups of definite

FIG. 354.

FIG. 353.

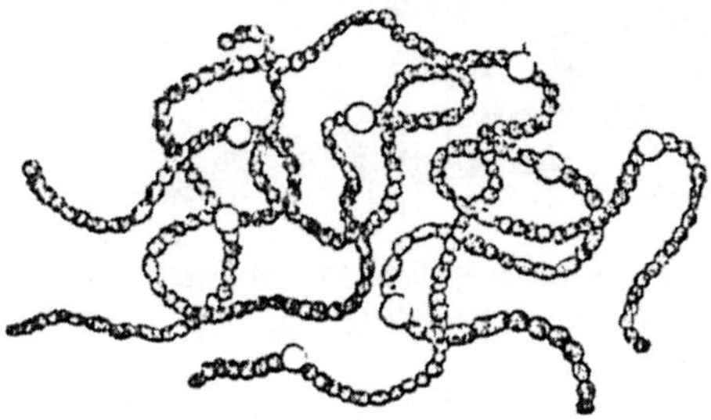

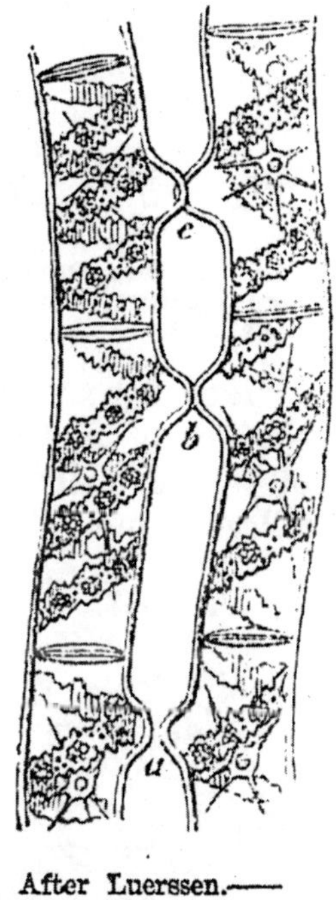

FIG. 355.

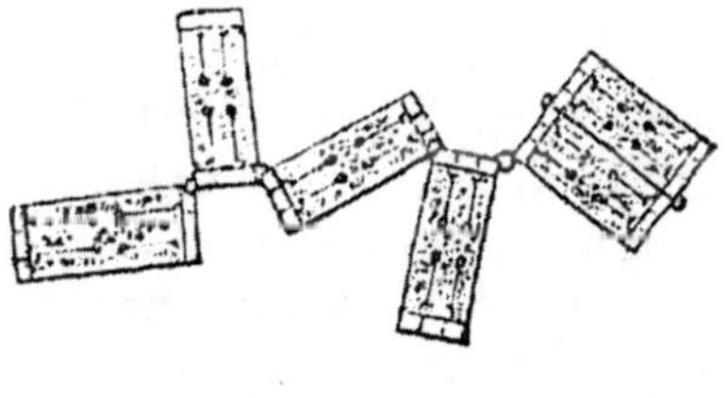

Fig. 353. Filaments from a *Nostoc* colony. After Luerssen.—*Fig.* 354. Two filaments of *Spirogyra* about to conjugate; each cell is seen to contain chlorophyll arranged in spiral bands with grains of starch, oil globules, and a central nucleus, surrounded by protoplasmic threads which extend to the cell-wall. *a, b, c.* Lateral protrusion of the cell-walls of adjoining cells.—*Fig.* 355. A species of Diatomaceous Alga (*Diatoma marinum*) divided into parts by *merismatic* or *fissiparous* cell-division. The parts are seen to be striated.

form, usually surrounded by a gelatinous investment, the cells exhibiting more or less regular geometrical outlines and enclosed by a membrane, striated (*fig.* 355) or granular, either simply tough and continuous (*fig.* 356), or impregnated with silex and separable into valves (*fig.* 355). Reproduction by spores formed after conjugation of the cells (*fig.* 356), by zoospores formed from the cell contents, and by division' (*fig.* 355). The Diatomaceæ are again

FIG. 356.

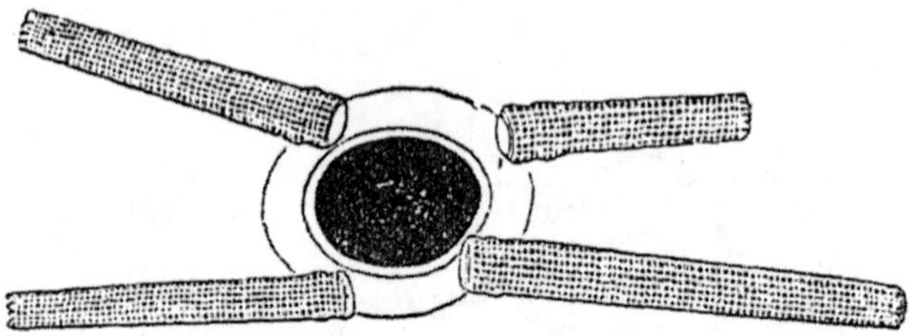

Fig. 356. Two Desmidiaceous Algæ (*Docidium Ehrenbergii*) after conjugation, with a *resting* or *inactive* spore between them. (After Ralfs.)

divided into two sections: 1. *Diatomeæ* (*fig.* 355). Natives of fresh or salt water, or of moist ground, of a brownish colour, valvular, and invested by a siliceous membrane. *Illustrative Genera*:—Diatoma, *DC.*; Navicula, *Bory*. 2. *Desmidieæ* (*fig.* 356). Found only in fresh water, of a green colour, continuous, containing starch, and not invested by a siliceous membrane. *Illustrative Genera*:—Closterium, *Nitzsch.*; Desmidium, *Ag.*

FIG. 357.

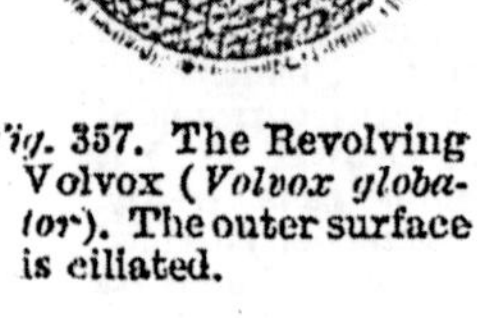

Fig. 357. The Revolving Volvox (*Volvox globator*). The outer surface is ciliated.

Sub-order 5. *Volvocineæ* (*fig.* 357).—Henfrey diagnoses them as follows:—'Microscopic bodies swimming in fresh water by the aid of cilia arranged in pairs upon the surface of a common semi-gelatinous envelope, the pairs of cilia each belonging to a green corpuscle resembling the zoospore of a confervoid, imbedded in the periphery of the common envelope. Reproduction by the development of each corpuscle into a new colony, the whole being set free by the solution of the parent envelope, or by conversion of the corpuscles into encysted resting-spores like those of Confervoids.' *Illustrative Genera*:—Volvox, *Lam.*; Gonium, *Lam.* The members of this group are frequently regarded as Infusorial Animalcules, but in all their essential characters they closely resemble the Confervoideæ; indeed, they are more commonly placed in this sub-order.

BOOK II.

DESCRIPTIVE BOTANY.

DESCRIPTIVE Botany is the art of describing plants in technical language, so that they may be readily recognised when met with by those to whom they were previously unknown. To describe a plant clearly and correctly necessarily implies a knowledge of the names of the different parts or organs of plants and of the various modifications which those organs present. Nothing, therefore, tends more to promote habits of observation in a student, and thus to induce a precise and practical knowledge of plants, than to cultivate the art of describing them; and it is, therefore, with a view of teaching students to describe correctly, and with precision and clearness, that the following brief instructions have been drawn up for their guidance.

CHAPTER I.

DIRECTIONS FOR DESCRIBING PLANTS.

Section 1. MEANS FOR OBSERVING PLANTS, AND GENERAL RULES FOR THEIR EXAMINATION.

FOR the study of plants, so far as is required in Descriptive Botany, all that is essentially necessary is a penknife or other sharp instrument, as a scalpel; a few common sewing needles, which may be mounted in handles made of pieces of wood or of ordinary pen-holders; a pair of small forceps; and an ordinary pocket lens, or a simple microscope. If an ordinary lens is used, this may be made, when necessary, into a simple dissecting microscope, by mounting it in such a manner, that it may be made to slide on an upright rod fixed below to a suitable stage upon which the object to be examined is laid, so that an observer may have the use of both hands when employed in its dissection, and care should be taken that the object is placed under the lens, and the light made to fall upon it. In examining minute objects a Coddington lens will also be found most useful, and may be mounted, when necessary, as in the case of the pocket lens just referred to, by suspending it by the loop with which it is provided to an upright rod, so that it may be moved up and down. Suitable stands for thus mounting lenses may be obtained of any optician; or a simple dissecting microscope may be at once obtained.

Other appliances that may be found necessary can be afterwards added according to the requirements of the student; and to those who can draw, a sketch-book and various coloured pencils will be found most useful, and should then always be employed to illustrate any points of structure not directly observable upon a superficial examination, such as dissections of certain parts of the flower, &c.

The student being furnished with the necessary appliances, should now provide himself with a specimen for examination. This should be carefully selected, and its characters closely observed and determined before they are written down—nothing should be glossed over. Another important circumstance also to be attended to in describing a plant, is to take the various parts in their proper order, as follows:—1. Root. 2. Stem. 3. Leaves. 4. Inflorescence. 5. Flower. 6. Fruit. 7. Seed. The separate portions of each part again, and of each organ, should, in like manner, be referred to in their order of development or arrangement. Thus, as regards the flower—1. Calyx; 2. Corolla; 3. Andrœcium; and 4. Gynœcium. And as to the Andrœcium—1. General characters of the stamens; 2. filament; and 3. anther; and then, as to the Gynœcium—1. General characters of the carpels; 2. ovary and ovules; 3. style; 4. stigma. And in like manner with the other organs.

Linnæus also insisted that in describing the parts and organs of a plant, these should be as clearly distinguishable one from the other as they are distinct in the plant itself. Hence, he required that every part and organ of a plant should have a paragraph to itself; and the student should, as a general rule, follow this instruction.

The following rule as to punctuation is also given by Lindley:—'In punctuating descriptions, always observe to separate—1. adjectives relating to the same noun by commas; 2. parts of the same organ by semicolons; 3. distinct organs by a period. Thus:—

'*Leaves* lanceolate, acuminate, villous; *petiole* short, winged, glandular; *stipules* rudimentary. *Racemes* terminal, nodding, many-flowered, secund.'

It will also be noticed that in the above illustration italics are employed for the organs and special parts to be described; and it is always important in these cases to use some means of marking such parts, so that they are readily distinguishable.

Section 2. Examination of the Special Organs and Parts of Plants.

In order to examine roots, stems, leaves, inflorescence, and their parts, little or no preparation is, as a general rule, required, because the characters of these are readily observable, and may be noted without dissection, and even, in most cases, without a lens, although the latter instrument is frequently useful, and in some instances indispensable, in the examination of the surface of such parts, and occasionally in other cases.

The following general directions may, however, be given for the examination of the different organs, &c. But as the terms employed in describing the various modifications to which they are liable are fully treated of in the author's 'Manual of Botany' and his 'Student's Guide to Structural, Morphological and Physiological Botany,' and generally also in works on Structural Botany, they require no notice here.

1. ROOT.—In examining what is supposed to be a root, the first point necessary is clearly to ascertain its nature, for if the presence of scales or leaf-buds can be demonstrated, it is an underground stem, and not a root, properly so called. When proved to be a root, the following points should be described in succession—its form, length, size, direction, ramification, colour, surface, texture, &c.

2. STEM.—The kind of stem should be first noticed, and properly designated as caulis, culm, rhizome, tuber, &c., and then further described as to its characters in the same order and in like manner as with the root.

3. LEAVES.—The position of these on the stem, as radical, cauline, &c., should be first noted; then in succession their arrangement, insertion, whether stipulate or exstipulate, direction, &c. The parts of the leaf should then be described. Thus, first the *lamina* in reference to its texture, colour, surface, venation, composition, margins, incision, general outline, form, and apex. The divisions of a simple leaf, and the leaflets of a compound leaf, should also be described in the same manner as the lamina itself; and in treating of compound leaves, it should be noted whether they are pinnate, ternate, bipinnate, &c.

The *petiole* should be noted as to its form, surface, length, &c.; and the *stipules*, being liable to similar modifications as regards margins, incision, outline, apex, &c., as the lamina, should be described in a similar manner when necessary, and the special kinds of stipules, as adnate, caulinary, or free, &c., should also be noted. In the case of leaf-buds, as also with flower-buds, it should be noticed whether they are axillary or otherwise, and described accordingly, and their mode of vernation or æstivation should be specially referred to. For this latter purpose, both in the case of leaf-buds and of flower-buds, vertical and horizontal sections will be necessary in order to determine the relation of the component organs to one another.

4. INFLORESCENCE.—Under this head we should notice whether *bracts* are present or absent, and then, when present, describe them in detail as with ordinary leaves, and the special varieties as involucre, spathe, &c., should be particularly noted. The *peduncle* should be described as to its form, direction, branching, surface, &c., and special kinds should be particularised. Then the nature of the inflorescence, whether indefinite or definite, and the particular kinds, as spike, raceme, umbel, &c., should receive marked attention. The position, direction, and number of flowers, &c., also require to be noticed.

5. FLOWER.—In examining the characters of flowers, great care should be taken in handling them so as not to crush or in any way injure the arrangement and relative position of their component parts. When the flowers are large they may be readily pulled to pieces, and most of their parts examined by means of the fingers, a scalpel or sharp penknife, and a needle mounted in a penholder. Great assistance will also be obtained by making perpendicular sections through the centre of entire flowers, and in order to avoid crushing or in other ways injuring the parts of the flower and their normal relations to one another, such sections should always be made from below upwards. Horizontal sections will also be frequently found very instructive.

especially as regards the central portions of the flower; and so far as flower-buds are concerned, as already noticed, they are indispensable.

When the flowers are of small size, and in all cases when examining the minute characters of ovules and seeds, an ordinary lens, mounted as already described, or a simple microscope, will be necessary. The parts of the flower, &c., can then be readily dissected on the stage below by a proper scalpel and needles, &c., and their characters noted.

In examining flowers, the CALYX should be first noticed in reference to its relative position as regards the ovary and other parts of the flowers, union or otherwise of its sepals, their number, shape, form, direction, colour, appendages, duration, &c.

The COROLLA should be treated as the calyx both as a collective organ and as regards its constituent petals.

The ANDRŒCIUM should now be studied in relation to the insertion, number, separation or union, and relative length of its stamens generally; and then the individual stamens should be referred to as regards their filaments and anthers. Thus, as to the *filament*, its form, length, colour, surface, direction, &c., should be noted; and then the *anther*, as to its attachment to the filament, and any appendages that may be present; its form, colour, mode of dehiscence, &c. The number of cells in the anther should also be carefully observed, and any peculiarities likewise of the *pollen* or *pollen grains* that can be noticed without the aid of a compound microscope, such as their separation or union, colour, &c.

Should a *disk* be present it must then be particularised, and attention should be especially directed to its insertion and general character.

Lastly, as regards the flower, the GYNŒCIUM will come under examination generally as follows — the number, union or separation, relative position to surrounding whorls, &c., of the constituent carpel or carpels; and then the individual carpels should be referred to as regards the ovary, style, and stigma. Thus as regards the *ovary*, its form, surface, number of cells, mode of placentation, &c.; and then the *styles* and *stigmas* must be observed. Thus the number, position, form, colour, surface &c. of these parts should be noted in due order. The number, position, mode of attachment, and relation of the parts to one another, of the ovules, must now be examined with great care, and in reference to many of these parts, as already mentioned, careful dissection and the aid of a lens will be necessary.

The *thalamus* may be either examined here or previously with the peduncle, and in doing so attention should be directed to its form, and any prolongations, such as carpophore, &c., must be especially noted.

6. FRUIT.—The general characters of the fruit should be described in the same manner as the gynœcium; but special attention should also be paid to its nature, mode of dehiscence, texture, and kind.

7. SEED.—*Seeds* must be examined in the same manner as *ovules*, as regards their number, mode of attachment, and position; and in a special manner in reference to their form, colour, surface, texture, appendages, and other marked characters furnished by the integuments; and afterwards the variations of the nucleus. Thus whether exalbuminous or albuminous, and in the latter case the kind and quantity of the albumen; and, lastly, the characters of the embryo should be as carefully ascertained as is possible, in reference to the number, form, size, and nature of its cotyledons, and their relation to each other and to the other parts of the embryo, and also the relation generally of the embryo to the other parts of the seed, and to the fruit.

Seeds with hard integuments or when dried should previous to their internal examination be soaked in warm water, for varying periods, but commonly for some hours, until their parts can readily be removed by dissection, and their relations to one another, &c. observed. In the case of small seeds, and those with smooth and slippery surfaces, they should be previously fixed in varnish or gum, so that they may readily be sliced, &c.

Section 3. Of the Characters of Plants.

Characters.—When a plant is described according to the rules laid down in the preceding section, it comprises all the characters of that plant, for by the term character we mean a list of all the points by which any particular *variety*, *species*, *sub-genus*, *genus*, *sub-tribe*, *tribe*, *sub-order*, *order*, *sub-class*, *class*, or *sub-kingdom* of plants, is distinguished from another. We have also two kinds of characters, which are called respectively *essential* and *natural*. By an essential character, we understand an enumeration of those points only by which any division of plants may be distinguished from others of the same nature; such may be called *diagnostic* characters. A *natural character*, on the other hand, is a complete description of a given species, genus, order, class, &c., including an account of every organ from the root upwards, through the stem, leaves, flowers, fruit, and seed. Such characters are necessarily of great length, and are not required for general diagnosis, although of great value when a complete history of a plant or group is required. Those characters, again, which refer to a species, are called *specific*, and are taken generally from all the organs of the plant, and relate chiefly to their *form*, *shape*, *surface*, *division*, *colour*, *dimension*, *duration*, &c., or, in other words, to characters of a superficial nature, and without reference to their internal structure. The characters of a genus are called *generic*, and are taken from the organs of reproduction. The characters of an order are termed *ordinal*, and are derived from the general structure of the plants in such groups, more especially of the organs of reproduction; while the characters of a class, as already mentioned, are derived from certain important structural peculiarities which the plants of such divisions exhibit. The essential character of a genus, when indicated in Latin, is put in the nominative case, while that of a species is placed in the ablative.

Abbreviations and Symbols.—It is usual in botanical works, when describing plants, to use certain abbreviations and symbols. A few of the more important need alone be mentioned here.

Abbreviations.—The names of authors, when of more than one syllable, are commonly abbreviated by writing the first letter or syllable, &c., as follows:—

L. or *Linn.* means Linnæus; *Juss.* is the abbreviation for Jussieu; *DC.* or *De Cand.* for De Candolle; *Br.* for Brown; *Lindl.* for Lindley; *Rich.* for Richard; *Willd.* for Willdenow; *Hook.* for Hooker; *With.* for Withering; *Endl.* for Endlicher; *Bab.* for Babington; *Berk.* for Berkley, &c., &c.

It is common to put such abridged names after that of the genus or species which has been described by them respectively. Thus *Eriocaulon, L.* indicates that the genus *Eriocaulon* was first described by Linnæus; *Miltonia, Lindl.* is the genus *Miltonia* as defined by Lindley; *Nuphar pumila, DC.* is the species of *Nuphar* defined by De Candolle, &c., &c.

Other abbreviations in common use are: *Rad.* for root; *Caul.* for stem; *Fl.* for flower; *Cal.* for Calyx; *Cor.* for corolla; *Per.* for perianth; *Fr.* for fruit; *Nat. Ord.* for natural order; *Gen.* for genus; *Sp.* or *Spec.* for species; *Var.* for variety; *Hab.* for habitat; *Herb.* for herbarium, &c. Again,—

V. v. c. (*Vidi vivam cultam*) indicates that the author has seen a living cultivated plant as described by him.

V. v. s. (*Vidi vivam spontaneam*) indicates that the author has seen a living wild plant.

V. s. c. (*Vidi siccam cultam*) indicates that a dried specimen of the cultivated plant has been examined.

V. s. s. (*Vidi siccam spontaneam*) indicates that a dried specimen of the wild plant has been examined.

Symbols.—The more important symbols are as follows:—

⊙, ○, ①, or A, signifies an annual plant.

⊙ ⊙, ②, or B, means a biennial plant.

♃, △, or P, signifies a perennial.

♄ or Sh. means a shrub.

T signifies a tree.

(twining to the right;) twining to the left.

♂ a staminate flower.

♀ a pistillate flower.

☿ an hermaphrodite flower.

♂ - ♀ a monœcious plant.

♂ : ♀ a diœcious species.

☿ ♂ ♀ a polygamous species.

○ = signifies that the cotyledons are accumbent, and the radicle lateral.

○ || Cotyledons incumbent, radicle dorsal.

○ ≫ Cotyledons conduplicate, radicle dorsal.

○ || || Cotyledons twice folded, radicle dorsal.

○ || || || Cotyledons three times folded, radicle dorsal.

? The note of interrogation is used to indicate doubt or uncertainty as to the genus, species, locality, &c.

! The note of exclamation indicates certainty in the above particulars.

* The asterisk indicates that a good description is to be found at the reference to which it is appended.

CHAPTER II.

EXAMPLES OF DESCRIPTIONS OF PLANTS.

THE examples of described plants given in this chapter have been selected from some of the more important natural orders, of which the characters have been given in the First Book of this volume—namely, that which relates to the Classification of Plants; and they will thus form not only general illustrations of the mode of describing plants, but also of the natural orders to which they belong, and of the divisions of plants. The descriptions are founded in nearly all cases upon those given in Bentley and Trimen's 'Medicinal Plants,' which were drawn up by Dr. Trimen, and a reference is always made to a figure to be found in that work. These descriptions have, however, been necessarily modified and shortened by the author of the present little work so as to adapt them to the special objects he had in view. The plants are all common in a wild or cultivated state in Britain, and may therefore be readily obtained for examination. They are, moreover, in most cases, important medicinal plants, and should be well known, at least by medical and pharmaceutical students. It is recommended that each specimen should at first be carefully examined and described by the student, as far as possible, in accordance with the rules laid down in the preceding chapter, and then the description thus made should be compared with that which is printed, and any important omission or error noted, and the specimen afterwards carefully re-examined in reference to such details. In this way the student will soon acquire a practical knowledge of the various organs and parts of plants, and be able to describe their more important characters clearly and correctly. He will also, after a time, find, as shown by the examples given, that while following in all essentials the rules laid down for his guidance, it is in some cases advisable, in the treatment of details and in their order of arrangement, to deviate in some degree from their rigid observance.[1]

[1] Blank forms for use in describing plants have been drawn up by Mr. E. M. Holmes, and will be found very convenient. They are published by Christy & Co., London.

ACONITUM NAPELLUS, *Linn.* Monkshood.

(Bentley and Trimen's *Medicinal Plants*, vol. i., plate 6.)

Class I. DICOTYLEDONES. Division I. ANGIOSPERMIA.

Sub-class 1. *Thalamifloræ. Nat. Ord.* Ranunculaceæ.

ROOT conical, some inches in length, tapering below to a fine point and giving off numerous slender branches in its course downwards, fleshy, dark brown or nearly black, smooth, and having at its upper part (one or more) short thick lateral shoots, each of which gives off at the end a new pale-coloured conical root or tuber, which is terminated above by a bud.

STEM erect, herbaceous, 2–4-feet high, unbranched, smooth below, slightly hairy above, green, bearing leaves below, and terminating above in the peduncle.

LEAVES alternate, exstipulate, on long petioles, which are dilated at the base, the upper ones sometimes sessile, spreading, palmatisected; *segments* 3 or 5, wedge-shaped, and deeply and irregularly divided into oblong or acute, narrow lobes, dark green, smooth, and shining above, paler beneath, and slightly hairy.

INFLORESCENCE indefinite, forming an erect, terminal raceme, flowers large and somewhat loosely arranged; *pedicels* erect, downy, in the axils of short lanceolate bracts, and with two smaller bracts (*bracteoles*) close to each flower.

CALYX inferior: *sepals* 5, petaloid, very unequal in size, deciduous, imbricate, dark bluish-purple; the upper or posterior one large, helmet-shaped, laterally compressed, pointed, and marked longitudinally with nearly parallel veins; the two lateral ones broadly ovate, blunt, hairy within, the two lower oblong or lanceolate, unequal.

COROLLA with 2–8 petals; petals unequal in size, the two posterior enclosed in the hood of the posterior sepal, arched and of very peculiar form, consisting of a long curved stalk, supporting at the end a small inflexed tongue-like portion, which is prolonged backwards into a rounded hollow knob, the remaining ones usually some (or all) abortive, very small, subulate, very similar to the filaments.

ANDRŒCIUM consisting of numerous hypogynous stamens; *filaments* rather long, dilated below, slightly hairy, the outer ones drooping; *anthers* adnate, small, roundish, 2-celled, dehiscing longitudinally, dull green.

GYNŒCIUM compound, apocarpous; *carpels* 3, shorter than the stamens, and somewhat divergent; *ovaries* superior, oblong, smooth, with numerous anatropous ovules in two rows; *styles* tapering; *stigmas* faintly bilobed.

FRUIT of 3 follicles; *follicles* erect, dry, papery, veined, scarcely an inch long, compressed-cylindrical, beaked by the recurved styles, dehiscing at the ventral suture.

SEEDS numerous, angular, irregularly pyramidal, brown; *testa* thick, and marked with irregular prominences and excavations; *embryo* very small at the base of abundant albumen.

PAPAVER RHŒAS, *Linn.* Red Poppy. Corn Poppy.

(Bentley and Trimen's *Medicinal Plants*, vol. i., plate 19.)

Class I. DICOTYLEDONES. Division I. ANGIOSPERMIA.

Sub-class 1. *Thalamifloræ.* *Nat. Ord.* Papaveraceæ.

ROOT tapering, somewhat branched.

STEM erect, herbaceous, 1–3 feet high, branched, round, green, rough with stiff spreading hairs or bristles.

LEAVES exstipulate; radical leaves stalked, obovate-lanceolate, pinnatisected, with lanceolate distant segments irregularily incised into ascending lobes tipped with bristles; stem leaves alternate, sessile or amplexicaul, triangular, pinnatisected with narrow irregularly incised lobes, bright green, hairy on both sides.

FLOWERS large, solitary at the ends of the stem and branches, on long peduncles, hispid with spreading or adpressed stiff hairs, ebracteated; *flower-buds* drooping.

CALYX inferior; *sepals* 2, which are deciduous, being pushed over the top of the flower as it expands, pale green, with spreading bristly hairs.

COROLLA, of four distinct petals; *petals* large, thin, deep scarlet or red, and often with a dark purple or almost black spot at the base, crumpled in æstivation; the two outer much wider than long, depressed; the two inner about as broad as long, concave, erect; all caducous.

ANDRŒCIUM of numerous hypogynous stamens; *filaments* filiform, blackish-purple; *anthers* innate, 2-celled, dehiscing longitudinally, small, violet.

GYNŒCIUM compound, syncarpous, of from about 10–14 carpels; *ovary* superior, roundish-obovoid, 1-celled, with 10 or more parietal placentas, which reach nearly to the centre; *ovules* numerous, anatropous; *stigmas* from about 10–14, sessile on and radiating from the centre of the summit of the ovary.

FRUIT a capsule, 1-celled, shortly stalked, roundish-obovoid, smooth, pale brown, surmounted at the flattish top by the radiating stigmas which project and form a crenated margin, dehiscing by pores beneath the stigmas.

SEEDS very numerous, small, attached all over the parietal placentas, reniform; *testa* reticulated; *embryo* minute, at the base of abundant fleshy-oily albumen.

BRASSICA NIGRA, *Koch*. Black Mustard.

(Bentley and Trimen's *Medicinal Plants*, vol. i., plate 22.)

Class I. DICOTYLEDONES. Division I. ANGIOSPERMIA.

Sub-class 1. *Thalamifloræ*. *Nat. Ord.* Cruciferæ.

ROOT small, irregularly branched.

STEM 2–4 feet high, herbaceous, stiff, with spreading

branches, dark green, more or less hispid below, usually glabrous above.

LEAVES alternate, stalked, exstipulate, dark green; lowest leaves lyrate, slightly hispid, with a large terminal lobe, intermediate ones irregularly pinnatifid, upper ones lanceolate, margins (except of the uppermost ones) faintly serrate.

INFLORESCENCE indefinite, corymbose at first, but ultimately racemose, ebracteated.

FLOWERS small; *pedicels* shorter than the calyx, stout, erect or nearly so.

CALYX of 4 distinct sepals, inferior, deciduous; *sepals* narrowly oblong, pale green; spreading almost horizontally.

COROLLA of 4 distinct petals, alternating with the sepals, deciduous; *petals* arranged in a cruciate manner, each with a narrow claw and roundish or somewhat obovate blade, bright yellow, spreading.

ANDRŒCIUM of 6 hypogynous stamens; *stamens* tetradynamous; the two with short filaments lateral, with a small gland placed above the base of each, the four with longer filaments placed in pairs antero-posteriorly, with a large gland at the base of each pair; *anthers* versatile, 2-celled, dehiscing longitudinally.

GYNŒCIUM compound, being formed of 2 united carpels; *ovary* superior, oblong-ovoid, tapering upwards into a very short style-like process, glabrous, 1-celled, with two parietal placentas; *ovules* several, stalked, campylotropous; *stigma* bilobed.

FRUIT a siliqua, about $\frac{3}{4}$ inch long, oblong-linear, adpressed, somewhat quadrangular, glabrous, tipped with a short tapering beak, 2-celled, dehiscing by 2 valves separating from below upwards, and each keeled by a single dorsal vein.

SEEDS stalked, 3–7 in each cell, arranged in a single row, but none in the beak, very small, roundish-oval; *testa* dark brownish-red, minutely pitted; *embryo* large, *cotyledons* conduplicate; exalbuminous.

ALTHÆA OFFICINALIS, *Linn.* Marsh Mallow.

(Bentley and Trimen's *Medicinal Plants*, vol. i., plate 35.)

Class I. DICOTYLEDONES. Division I. ANGIOSPERMIA.

Sub-class 1. *Thalamifloræ.* *Nat. Ord.* Malvaceæ.

ROOT tapering downwards, large, thick, yellowish-white, tough, fleshy, a foot or more in length.

STEMS several, herbaceous, erect, stiff, 2–4 feet high, unbranched or nearly so, cylindrical, and covered with a very dense velvety pubescence of stellate hairs.

LEAVES alternate, stipulate, stalked; the lower ones roundish-ovate, the upper triangular-oval or somewhat 3–5-lobed, irregularly cut, acute, coarsely serrate, somewhat plaited, velvety on both sides with short close hairs, pale greyish green; *stipules* caulinary, narrowly triangular, caducous.

FLOWERS in small axillary cymose clusters, which are shorter than the leaves, or rarely solitary; each flower shortly stalked and surrounded by an involucre or epicalyx, with from 7–10 subulate-triangular erect segments.

CALYX inferior, gamosepalous, divided into 5 narrowly triangular acute segments, densely pubescent, persistent; valvate in æstivation.

COROLLA composed of 5 obcordate wedge-shaped petals united by their narrow bases and adherent to the tube formed by the coherent filaments, pale purplish-rose coloured; twisted in æstivation.

ANDRŒCIUM composed of numerous stamens, which are united below by their filaments into a column or hollow tube, which is attached at its base to the corolla; *anthers* drooping, reniform, 1-celled, dark purple, dehiscing transversely.

GYNŒCIUM compound, and composed of several carpels which are closely arranged so as to form by their ovaries, which are superior, a roundish flattened whorl, which is covered by the base of the combined petals and staminal tube; *ovules* solitary in each ovary or cell, amphitropous; *styles* united below into a cylinder, which passes through the staminal tube, and divides above into numerous filiform branches, with the stigmatic surfaces on their inside.

FRUIT brownish green, flattened, partially covered by the persistent calyx, and consisting of numerous dry laterally compressed indehiscent carpels, arranged in a whorl and attached to a short central axis, and forming together the kind of fruit known as a carcerule; each carpel hairy on the back, and separating from each other in a septicidal manner when ripe.

SEEDS in each carpel solitary, ascending, reniform, smooth, brown; *embryo* curved with thin cotyledons; *albumen* nearly absent.

CYTISUS SCOPARIUS, *Link*. Broom.

(Bentley and Trimen's *Medicinal Plants*, vol. ii., plate 70.)

Class I. DICOTYLEDONES. Division I. ANGIOSPERMIA.

Sub-class 2. *Calyciflorœ*. 1. *Perigynœ*.

Nat. Ord. Leguminosæ.

ROOT woody, long, irregularly branched.

STEM short, much branched, forming a bush of 4–5 feet in height; *branches* erect, angular, furrowed, with dark yellowish-green bark, and long wand-like twigs; *twigs* tough, blunt at the ends, which are downy, dark green, angular, and with 5 prominent leaf-like wings.

LEAVES alternate, exstipulate or the stipules minute; the lower ones on flat short petioles, trifoliate, the upper ones sessile, unifoliate, leaflets sessile, lanceolate or somewhat obovate, sub-acute, entire, dark green and smooth or slightly hairy above, hairy below.

INFLORESCENCE indefinite, racemose.

FLOWERS numerous, large, solitary, on smooth pedicels of varying lengths in the axils of the sessile leaves.

CALYX inferior, gamosepalous, compressed laterally, smooth, and divided into two brown scarious shallow lips, the upper minutely 2- the lower 3- toothed, persistent.

COROLLA of 5 distinct spreading petals, bright yellow, arranged in a papilionaceous manner; *standard* roundish, with a short claw, notched, with somewhat involute margins, *wings* oblong, *keel* obtuse, as long as wings, ultimately deflexed.

ANDRŒCIUM of 10 stamens, which are inserted at the base of the calyx-tube; *filaments* all united for some distance, 5 longer than the other 5, and alternating with them, curved upwards; *anthers* small, the five on short filaments being versatile, and the five on long filaments somewhat inuate.

GYNŒCIUM simple; *ovary* sessile, superior, compressed laterally, with dense silky hairs along each margin; *ovules* many, attached to the ventral suture; *style* long, much curved or even coiled into a ring, smooth; *stigma* terminal.

FRUIT a legume, about 1½ inches long, oblong or linear-oblong, curved or nearly straight, apicilar, much flattened, dark brown, smooth, except at the margins, where there is a fringe of long white hairs, 2-valved, the valves separating with elasticity and becoming afterwards much twisted.

SEEDS from 12–18 in each legume, each with a short funiculus, which is swollen at the hilum, oblong-ovoid, smooth and shining, hard, olive-coloured; *radicle* thick, incumbent; *albumen* in small quantity.

POTENTILLA TORMENTILLA, *Stokes*. Tormentil.

(Bentley and Trimen's *Medicinal Plants*, vol. ii, plate 101.)

Class I. DICOTYLEDONES. Division I. ANGIOSPERMIA.

Sub-class 2. *Calycifloræ*. 1. *Perigynæ*.

Nat. Ord. Rosaceæ.

RHIZOME perennial, short, nearly cylindrical, branched, giving off below long cylindrical rootlets, and reddish-brown chaffy scales above, externally dark brown, and internally bright red or some tint of red.

STEMS numerous, herbaceous, from the axils of the chaffy scales, very slender, cylindrical, prostrate or ascending, 1 foot or more long, pale green or reddish, slightly hairy, much branched above.

LEAVES stipulate; radical leaves alternate, shortly stalked, and usually ternate or rarely quinate, the cauline ones generally opposite, nearly or quite sessile, and trifoliate; leaflets of all small, sessile, obovate or lanceolate-wedge-shaped, entire and tapering below, and with a few large serratures or teeth above, hairy on both surfaces, dark green above, paler below; *stipules* lanceolate or oval, entire or palmately toothed or incised.

FLOWERS small, cymose, on long slender stalks terminating the stems and branches, and each flower surrounded by an involucre, which is deeply cut into 4 somewhat linear or lanceolate segments.

CALYX inferior, gamosepalous, with a perigynous hairy rim or disk at its base, deeply cut into usually 4 lanceolate or somewhat ovate, acute, pale green, hairy, valvate segments, which alternate with the shorter segments of the involucre, persistent.

COROLLA of usually 4 distinct petals; *petals* roundish or obovate, with very short claws, perigynous, spreading, yellow, and usually darker at the base, deciduous.

ANDRŒCIUM composed of about 16 stamens, which are attached to the perigynous hairy disk; *filaments* slender, shorter than the petals, distinct, yellowish; *anthers* short, innate, rounded.

GYNŒCIUM compound, and formed of about 6–8 carpels, which are apocarpous and arranged on a flattish dry thalamus; *ovaries* superior, small, glabrous; *styles* lateral, erect, yellow; *ovules* solitary.

FRUIT consisting of a few achenia, each of which is somewhat reniform, keeled on the back, brown, smooth when fresh, but when dry reticulated.

SEEDS solitary, pendulous, exalbuminous; *radicle* superior.

CONIUM MACULATUM, *Linn.* Hemlock.

(Bentley and Trimen's *Medicinal Plants*, vol. ii., plate 118.)

Class I. DICOTYLEDONES. Division I. ANGIOSPERMIA.

Sub-class 2. *Calyciflorœ.* 2. *Epigynœ.*

Nat. Ord. Umbelliferæ.

ROOT long, forked, pale yellow, half an inch or more in diameter.

STEM herbaceous, solitary, erect, 3–6 feet high, stout below, much branched above, striated, hollow, not enlarged in a marked manner at the nodes, perfectly smooth and covered with a whitish bloom which is very easily rubbed off, bright green mottled with irregular stains or spots of a purplish colour.

LEAVES exstipulate, numerous; the lower ones alternate, very large, sometimes as much as 2 feet, deltoid in outline, with long amplexicaul petioles, usually decompound or tripinnate; the upper ones much smaller, nearly sessile, with short amplexicaul petioles, often opposite or three together, bipinnate or pinnate. All dull green, glabrous, and the ultimate divisions lanceolate-oblong, pinnatisected, with toothed segments, each tooth tipped with a minute, acute, white point.

INFLORESCENCE consisting of numerous terminal compound umbels with flattened tops. *Umbels* composed of from about 12–16 straight primary rays, and situated on rather short peduncles and each umbel surrounded by both a general and partial involucre; the *general* of about 4–8 lanceolate, acuminate, deflexed bracts bordered with white, and the *involucels* of 3 or 4 small spreading bractlets, which are unilateral or arise from the outer side only.

FLOWERS small, white, the outer ones usually rather the larger, and the inner ones often barren.

CALYX completely adherent to the ovary, the limb only forming a thick epigynous ring.

COROLLA of 5 distinct petals, epigynous; *petals* usually nearly equal, obovate, inflexed at their points.

ANDRŒCIUM of 5 distinct stamens, epigynous, alternate with the petals, incurved; *anthers* white.

GYNŒCIUM compound, of two united carpels; *ovary* inferior, 2-celled, crowned with a double flattened white fleshy disk (stylopod) and distinctly marked with faintly crenated primary ridges; *styles* 2, short, reflexed; *ovules* solitary in each cell, pendulous.

FRUIT a cremocarp, small, about $\frac{1}{8}$ inch long, broadly ovoid, somewhat compressed laterally, crowned by the dry stylopod and 2 reflexed slender styles, smooth; *mericarps* readily separating from an undivided carpophore, primary ridges prominent, equal, more or less crenated, the two lateral marginal, without manifest vittæ, but the channels marked at first by numerous slender irregular streaks.

SEEDS solitary in each mericarp, pendulous, deeply grooved ventrally; *embryo* minute, at the base of abundant horny albumen; *radicle* towards the hilum.

MATRICARIA CHAMOMILLA, *Linn.* Wild Chamomile.

(Bentley and Trimen's *Medicinal Plants*, vol. iii., plate 155.)

Class I. DICOTYLEDONES. Division I. ANGIOSPERMIA.

Sub-class 3. *Corollifloræ.* 1. *Epigynæ.*

Nat. Ord. Compositæ.

ROOT fibrous.

STEM herbaceous, erect, 1–2 feet high, much branched, solid, smooth, strongly striated, pale green; *branches* long and slender.

LEAVES exstipulate, numerous, alternate, sessile, amplexicaul, bi- or tripinnatisected, the ultimate segments narrow, setaceous, acute, spreading, smooth, bright green.

INFLORESCENCE consisting of numerous capitula, which are placed at the ends of the branches, the whole being arranged in a somewhat corymbose manner; the separate capitula about $\frac{5}{8}$ of an inch wide, and surrounded by a flat involucre composed of 1, 2, or 3 rows of very small, equal, linear, smooth, obtuse bracts (phyllaries) with scarious brownish ends and transparent margins; *receptacle* broadly ovoid and solid at first, but ultimately becoming conical and hollow, smooth and without scales.

FLOWERS OF FLORETS yellow in the centre or *disk* of the capitula; and white at the *ray* or circumference.

DISK FLORETS small, numerous, bisexual, each with a superior *calyx* without any evident limb, and a deeply 5-toothed tubular *corolla*, which is valvate in æstivation. *Andrœcium* of 5 stamens with syngenesious anthers, each of which has an appendage at the apex; *gynœcium* compound, syncarpous; *ovary* inferior, 1-celled, terminated by a solitary cylindrical style, which is bifid above and each arm stigmatic at its apex, and thus showing that the gynœcium is formed of two united carpels; *ovule* solitary, erect.

RAY FLORETS varying in number from about 15–25, unisexual, female, crowded and overlapping, each with a superior *calyx* without any evident limb; a ligulate oval-oblong *corolla*, which is faintly and bluntly 2–3-lobed, and spreading when in full flower, but afterwards deflexed; *style* cylindrical, with spreading arms.

FRUIT a cypsela, very small, oblong-ovoid, somewhat curved, with 5 slender ribs on its ventral or concave face, quite smooth, pale grey, and crowned by a very slight somewhat membranous border, but no pappus, 1-celled.

SEED solitary, erect, exalbuminous; *radicle* inferior.

ARCTOSTAPHYLOS UVA-URSI, *Spreng.* Bearberry.

(Bentley and Trimen's *Medicinal Plants*, vol. iii., plate 163.)

Class I. DICOTYLEDONES. Division I. ANGIOSPERMIA.

Sub-class 3. *Corolliflorœ.* 2. *Hypostamineœ.*

Nat. Ord. Ericaceæ.

STEM woody, decumbent, rooting, much branched, the branches turning upwards, somewhat downy when young; bark ultimately brownish and scaling.

LEAVES numerous, alternate, crowded, shortly stalked, exstipulate, evergreen, $\frac{1}{2}$ to 1 inch long, thick and coriaceous, obovate or spathulate, obtuse, tapering at the base, entire with the margins somewhat rounded, young leaves minutely pubescent, but ultimately smooth on both surfaces, shining and dark green above, paler and reticulated beneath; *petiole* somewhat woolly.

INFLORESCENCE in terminal, short, crowded, drooping few-flowered racemes or clusters; *pedicels* short, thick, glabrous; *bracts* very short, acute, thin.

CALYX inferior, small, divided into 4 or 5 broad, roundish, smooth, thin, pinkish segments, with ciliated margins, persistent.

COROLLA gamopetalous, urceolate, about $\frac{1}{3}$ of an inch long, with 4 or 5 small, triangular, spreading teeth, smooth outside, hairy within, pale pinkish white, with somewhat rose-coloured teeth.

ANDRŒCIUM of 8 or 10 stamens, distinct, hypogynous or slightly adherent to the base of the corolla; *filaments* very short, somewhat flattened, curved inwards, hairy below; *anthers* 2-celled, deflexed, ovoid, each cell with a large pore at the apex, and a long spreading or somewhat reflexed awn.

GYNŒCIUM compound, syncarpous; *ovary* superior, roundish, fleshy, 5-celled, with a solitary ovule in each cell, surrounded by a shallow annular hypogynous disk, with 8 or 10 thickened blunt lobes, which alternate with the stamens; *style* simple, thick, much longer than the stamens; *stigma* terminal, simple, obtuse.

FRUIT small, globular, succulent, smooth, bright red, with a thick epicarp, and about 5 seeds.

SEEDS pendulous, shortly stalked, somewhat angular; *embryo* minute, straight, in the axis of fleshy albumen.

HYOSCYAMUS NIGER, *Linn.* Henbane.

(Bentley and Trimen's *Medicinal Plants*, vol. iii., plate 194.)

Class I. DICOTYLEDONES. Division I. ANGIOSPERMIA.

Sub-class 3. *Corolliforæ.* 3. *Epipetalæ.*

Nat. Ord. Solanaceæ.

ROOT long, tapering, irregularly branched, wrinkled, brown.

STEM herbaceous, erect, 1—2 feet high, much branched, cylindrical, and covered with long pale viscid soft glandular hairs.

LEAVES exstipulate, simple; radical leaves large, 6—8 inches or more long, stalked, triangular-ovate, with undulated sinuated margins; stem-leaves much smaller, alternate, sessile, more or less amplexicaul, oblong or ovate-oblong, acute, and irregularly toothed, sinuated, or somewhat pinnatifid; all pale green, thin, slightly hairy above, more so on their lower surface, and the veins likewise covered with glandular hairs like those of the stem.

INFLORESCENCE consisting of numerous sessile or shortly-stalked, crowded flowers, which are solitary in the axils of large spreading bracts, which resemble the leaves in appearance, the whole forming a kind of scorpioid spike or raceme, which elongates and straightens out after flowering.

CALYX gamosepalous, inferior, persistent. large, somewhat urceolate or bell-shaped, with a spreading limb with 5 shallow, equal, acute, ovate or broadly-triangular, acute teeth or lobes, the whole covered with long viscid glandular hairs.

COROLLA gamopetalous, with a funnel-shaped tube and spreading limb; *lobes* 5, broad, rounded, obtuse, slightly unequal, imbricated in the bud, lurid-yellow or straw-coloured, veined with purple, and with a purple throat.

ANDRŒCIUM consisting of 5 distinct stamens, which are inserted near the base of the tube of the corolla, but remaining attached to about the middle, declinate; *filaments* unequal in length, shorter than the corolla, and somewhat hairy; *anthers* dorsifixed or adnate, purple, 2-celled, dehiscing longitudinally.

GYNŒCIUM compound, of 2 united carpels; ovary superior, roundish, enclosed in the tube of the calyx, smooth, 2-celled, with large axile placentas; *ovules* numerous; *style* declinate, filiform; *stigma* capitate.

FRUIT enclosed in the enlarged persistent globular veined tube of the calyx, dehiscing transversely, and thus forming the kind of capsule which is called a pyxis, 2-celled.

SEEDS very numerous, somewhat reniform, brown, reticulated; *embryo* much curved in albumen.

DIGITALIS PURPUREA, *Linn.* Foxglove.

(Bentley and Trimen's *Medicinal Plants*, vol. iii., plate 195.)

Class I. DICOTYLEDONES. Division I. ANGIOSPERMIA.

Sub-class 3. *Corolliflorœ.* 3. *Epipetalœ.*

Nat. Ord. Scrophulariaceæ.

ROOT consisting of numerous long thick rootlets.

STEM solitary or several, herbaceous, erect, 2—7 feet or more high, simple or slightly branched, downy.

LEAVES exstipulate, alternate, simple, the lower ones large, often a foot or more, and tapering at the base into a long winged petiole, ovate-oblong or ovate-lanceolate, s... or irregularly crenate-dentate or serrate, crenate or ... somewhat rugose and slightly hairy on their upper su... densely pubescent beneath; the upper leaves becom... gradually narrower and more shortly-stalked, and at ... th passing into the sessile bracts.

INFLORESCENCE indefinite, ... forming an elongated terminal raceme, with leafy bra...; *flowers* very numerous, large, closely-packed, overlapping, pendulous, unilateral, the rachis and pedicels downy.

CALYX gamosepalous, inferior, persistent, very deeply divided into 5 acute oblong-lanceolate spreading segments of unequal size, finely downy.

COROLLA gamopetalous, $1\frac{1}{2}$—$2\frac{1}{2}$ inches long, irregularly bell-shaped, obscurely 2-lipped, the upper lip entire or divided into two faint lobes, the lower 3-lobed, with the central one much the largest, smooth, and crimson outside, paler-coloured within; the lower lip furnished with long hairs inside, and marked with dark crimson spots, each surrounded with a white border.

ANDRŒCIUM of 4 stamens, didynamous, arising from the base of the corolla, included; *filaments* thick, the two lower longer than the upper, kneed; *anthers*, 2-celled, cells diverging.

GYNŒCIUM compound, syncarpous; *ovary* superior, 2-celled, with numerous ovules in each cell, arising from large spongy axile placentas; *style* filiform, bifid above, and forming two stigmas.

FRUIT an ovoid capsule, tapering at the apex, 2-celled, 2-valved, dehiscing septicidally, and the valves also separating from the axis; *pericarp* thin and papery.

SEEDS very numerous, minute, oblong or ovoid, light brown, pitted; *embryo* straight, in the axis of albumen.

...ort or Snakeweed.

POLYGONUM BISTORTA, *Linn.*

(Bentley and Trimen's *Medici*... *ants*, vol. iii., plate 212.)

Class I. DICOTYLEDONES. ...Division I. ANGIOSPERMIA.

Sub-class IV. ...ochlamydeæ or *Incompletæ*.

... *Ord.* Polygonaceæ.

RHIZOME cylindrical, usually twice twisted so as to resemble the letter S in form, much branched, creeping, woody, brownish and marked with leaf-scars externally, pale red within, and giving off numerous fibrous roots.

STEM solitary, herbaceous, erect, simple, 1½—2 feet high, round, smooth, striated.

LEAVES alternate, stipulate; radical leaves ovate, acute, on long winged petioles, and with entire somewhat undulated margins; stem leaves few, ovate-lanceolate, acute, rapidly decreasing in size upwards, nearly sessile; all dark green and smooth above, glaucous beneath, except being pubescent on the prominent veins; *stipules* ochreaceous, truncate, membranous above.

INFLORESCENCE indefinite, forming a dense, terminal, erect, cylindrical, many-flowered raceme, 1—2 inches long.

FLOWERS in pairs, on stalked slender pedicels, one much more expanded than the other, and each pair surrounded at the base by two membranous cuspidate bracts.

CALYX inferior, composed of 5 oval, obtuse, smooth, pale pink or flesh-coloured sepals, which are slightly connected at the base, imbricate.

ANDRŒCIUM of 8 stamens, which are exserted, and hypogynous or somewhat attached to the base of the calyx; *filaments* distinct, slender, white; *anthers* small, versatile, purple.

Gynœcium compound and syncarpous; ovary solitary, small, superior, trigonous; *styles* 3, distinct, thread-like, shorter than the stamens; *stigmas* capitate.

Fruit small, dark brown, smooth, shining, acute, trigonous, indehiscent.

Seed solitary, erect, completely filling the pericarp; *embryo* lateral, curved, with a superior radicle; *albumen* mealy.

Salix alba, *Linn.* White Willow.

(Bentley and Trimen's *Medicinal Plants*, vol. iv., plate 234.)

Class I. Dicotyledones. Division I. Angiospermia.

Sub-class IV. *Monochlamydeæ or Incompletæ.*

Nat. Ord. Salicaceæ.

Root large, and much branched irregularly.

Stem tall, sometimes as much as 60 feet or more high, and 20 feet in circumference, covered with a thick, grey or yellow, furrowed, fissured bark; *branches* numerous, spreading, more or less ascending, the young shoots and buds covered with silky hairs.

Leaves numerous, alternate, stalked, stipulate; *blade* simple, 3—4 inches long, lanceolate or oval-lanceolate, much tapering, acute, glandular-serrate, whitish with adpressed silky hairs, more especially on the under surface, sometimes glabrous when old on the upper surface, midrib prominent beneath; *petiole* short, silky; *stipules* small, erect, lanceolate or subulate, very deciduous.

Inflorescence indefinite, consisting of very small, numerous, unisexual, diœcious, sessile flowers; each axillary to a small, oval-oblong or linear, sub-acute, ciliated, scaly bract, the whole forming a compact, slender, cylindrical, erect or spreading amentum from $1\frac{1}{2}$—$2\frac{1}{2}$ inches long; the several amenta placed on short, lateral, leafy stalks.

Male Flowers achlamydeous, but furnished with

two very small glandular scales, forming a kind of disk at the base of the stamens. *Andrœcium* of two distinct stamens; *filaments* rather long, hairy below, stiff; *anthers* small, roundish, yellow, 2-celled.

FEMALE FLOWERS achlamydeous, scales as in the male flowers. *Gynœcium* compound, of two united carpels; *ovary* sessile, ovate, tapering, smooth, 1-celled; *ovules* numerous, parietal; *style* very short; *stigmas* 2, spreading, each 2-lobed.

FRUIT a capsule, small, dry, conical, dehiscing loculicidally into 2 recurved valves, 1-celled.

SEEDS numerous, very small, each enclosed in long silky hairs, which arise below from a very short stalk or funiculus; *embryo* with an inferior radicle, and flat oval cotyledons; *albumen* none.

JUNIPERUS COMMUNIS, *Linn.* Juniper.

(Bentley and Trimen's *Medicinal Plants*, vol. iv., plate 255.)

Class I. DICOTYLEDONES. Division II. GYMNOSPERMIA.

Nat. Ord. Coniferæ.

STEM much branched, so as to form a bushy evergreen shrub of from 2—5 feet high, or in some cases a small tree as much as 30 feet high; *bark* reddish or chocolate-brown, flaking and fibrous on the old branches.

LEAVES very numerous, sessile, persistent, exstipulate, usually in threes, spreading at right angles, rigid, linear-acerose, channeled above and glaucous or pinkish with green margins; bluntly keeled and bright green beneath.

FLOWERS diœcious, axillary. *Barren* or *male flowers* in globose or somewhat ovoid catkins, about $\frac{1}{8}$ inch long, having a few small scales at the base; *anthers* in several whorls, each 3—6-celled, attached to the base of the lower surface of the overhanging connective,

dehiscing longitudinally; *pollen* spheroidal. *Fertile* or *female flowers* in cones about $\frac{1}{8}$ inch long, and consisting of from about 12—16 green, ovate, acute, fleshy scales, closely packed in 3—5 whorls on a short axis, and overlapping; the lower empty, and the uppermost three thinner, and alternating with 3 erect, naked ovules; *ovules* flask-shaped, with prolonged micropyles.

FRUIT a galbulus, formed from the three uppermost scales of the cone which have become much enlarged and fleshy, and entirely united except at their ends, where they form three radiating furrows, green at first, and not ripening till the second year, when the fruit is round and about the size of a pea, very fleshy, purplish-black, and covered with a blue-grey bloom, each fruit surrounded at the base by the stellately arranged dry empty scales of the cone.

SEEDS 3 in each fruit, lying close together, each seed somewhat triangular and sharp-edged at its upper extremity, and rounded below; *testa* very hard, and bearing below on its inner and outer surfaces, but more especially on the outer, several evident oil-glands; *albumen* in small quantity; *embryo* straight, with 2 oblong cotyledons, and a superior radicle.

ORCHIS MA'CULATA, *Linn.* Spotted Hand-Orchis.

Class II. MONOCOTYLEDONES.

Sub-class 1. *Petaloideæ.* 1. *Epigynæ.*

Nat. Ord. Orchidàceæ.

ROOT of two flattened palmate tubercules, and a few filiform unbranched rootlets.

STEM herbaceous, solitary, erect, from 6 inches to a foot or more high, leafy, solid.

LEAVES radical and cauline, simple, with parallel veins, usually with purplish-black spots, or rarely without spots, sheathing, acute or obtuse, glabrous, entire; the radical ones oblong-lanceolate or lanceolate-obovate, those of the stem alternate, oblong-lanceolate below, narrowing above and becoming linear-lanceolate so as to resemble the bracts.

FLOWERS in a dense terminal spike, which is at first short and somewhat ovate, but ultimately elongate and oblong-pyramidal, bracteate, usually very pale purple and spotted, or rarely white; *bracts* green, 3-veined, the lateral veins indistinct, linear-lanceolate, acute, the lower ones longer than the ovary, the upper about the same length.

PERIANTH superior, irregular, of six segments; three outer or *sepals*, spreading, oblong-lanceolate, lateral. oblique; the three inner or *petals*, unequal, the two lateral ovate, obtuse, and converging, the lower forming the *lip* or *labellum*; *lip* spurred, usually flat or sometimes reflexed at the margins, about as broad as long, deeply 3-lobed, lateral lobes somewhat rounded and crenulate, the middle lobe narrower, ovate or oblong, obtuse, and usually somewhat longer than the lateral petals, *spur* straight, subulate, and usually shorter than the ovary.

ANDRŒCIUM confluent with the style into a short column or gynostemium; *anther* 1, erect, 2-celled, dehiscing longitudinally, *pollen* cohering so as to form 2 *pollen-masses* (pollinia), each of which has a caudicle or stalk, which is imbedded at the period of dehiscence in the retinacula, which are placed on a beak-like minute projection from the style (*rostellum*) situated at the base of the anther.

OVARY inferior, twisted, 1-celled; *ovules* very numerous, attached parietally in 3 rows, anatropous; *style* forming an irregular column with the andrœcium; *stigma* concave, beneath the rostellum.

FRUIT capsular, 3-valved, the valves separating from a framework bearing the placentas, and adhering at the base and apex.

SEEDS parietal, very numerous, minute; *testa* loose, netted; *embryo* solid, fleshy; *albumen* absent.

Colchicum autumnale. Meadow Saffron.

Bentley and Trimen's *Medicinal Plants*, vol. iv., plate 288.)

Class II. Monocotyledones.

Sub-class 1. *Petaloideæ.* 2. *Hypogynæ.*

Nat. Ord. Melanthaçeæ or Colchicaceæ.

Stem.—No aerial stem, but a subterranean corm, from which a tuft of filiform roots arises beneath; *corm* (in autumn) 1½ inches or more in length, and 1 inch or more in diameter, irregularly pear-shaped, oblique at the base, covered by a shining dark chestnut-brown coat, and crowned by the withered remains of the leaves or with a hollow scar, rounded on one surface, and flattened on the other, the latter having attached at the base a very small new corm, from which the flower arises.

Leaves radical, appearing in the spring and withering before the autumn, 3–5, strongly sheathing, 6–12 inches long, erect, oval-lanceolate, entire, smooth, dark green.

Flowers large, appearing in the autumn of the year preceding the spring in which the leaves appear, solitary or 2 or 3, nearly sessile on the top of the new corm, erect, and surrounded at their lower part by a white spathaceous *bract*; and enclosed in the coats on the flattened side of the old corm.

Perianth inferior, with a very long slender tube; *limb* somewhat campanulate, pale purple, with 6 oblong segments arranged in two series, the inner a little smaller.

Andrœcium of 6 stamens inserted in the mouth of the tube of the perianth, and much shorter than the segments; *filaments* thread-like; *anthers* yellow, oblong-linear, attached at first above their base in front, ultimately versatile, 2-celled, bursting laterally.

Gynœcium compound, and formed of 3 carpels, which are united below so as to form a 3-celled superior *ovary*, which is subterranean at the bottom of the tube of the perianth; *ovules* numerous on axile placentas; *styles* 3, quite distinct, filiform, and so long as to project beyond the stamens, tips purple, recurved, and stigmatic on their inner surfaces above.

Fruit a septicidal capsule, 3-celled, 3-valved, membranous, slightly rugose, pale-brown, somewhat acute at both ends, but more especially at the upper.

Seeds numerous, roundish, with a short, fleshy funiculus; *testa* thick, rugose, brown; *embryo* minute, excentric; *albumen* abundant.

Triticum sativum, *Lam.* Wheat.

(Bentley and Trimen's *Medicinal Plants*, vol. iv., plate 294.)

Class II. Monocotyledones.

Sub-class 2. *Glumaceæ.* *Nat. Ord.* Graminaceæ.

Root consisting of many long much-branched rootlets, annual.

Stems several, herbaceous, erect, cylindrical, jointed, striated, glabrous, hollow except at the nodes, which are swollen.

Leaves alternate, few, distant, sheathing; *sheath* long, striated, round, usually smooth above, and rough with short deflexed hairs below; *ligule* very short, truncate or torn, membranous; *blade* simple, with parallel veins, 6–18 inches long, linear, acuminate, glabrous on both surfaces, or somewhat scabrous above.

Inflorescence indefinite, of from 12–24 sessile spikelets, arranged in a distichous manner on alternate surfaces of the flattened, deeply-excavated rachis, with a solitary terminal one, and forming together an oblong-linear, acute, terminal, tetragonal spike, from

3–5 inches long. *Spikelets* compressed, 3- or more flowered, with the terminal flower barren, and each with 2 glumes; *glumes* nearly equal, irregularly boat-shaped, oblong-oval or ovate, obtuse, apiculate, stiff, usually smooth and shining except at the midrib which is somewhat prickly.

EACH FERTILE FLOWER with 2 pales, 2 lodicules, 3 stamens, and a gynœcium of two united carpels. *The outer lower pale or flowering glume* boat-shaped, keeled above, ovate-oblong, obtuse or mucronate, or more or less awned at the apex, obscurely veined, smooth or hairy, coriaceous; *the upper pale* equalling the flowering glume, oval, obtuse, thin, papery and transparent, with two sharply-prominent green veins rough with minute prickles, and strongly inflexed margins. *Lodicules* thick, usually rounded and entire, hairy at the top.

ANDRŒCIUM of 3 hypogynous stamens; *filaments* long, capillary; *anthers* versatile, large, oblong-linear, very pale yellow, 2-celled, dehiscing longitudinally, and hanging out of the flowers after dehiscence.

GYNŒCIUM compound, syncarpous; *ovary* superior, obovate or globose, tufted with white hairs at the top, 1-celled; *ovule* solitary; *styles* 2; *stigmas* feathery.

FRUIT a caryopsis, enclosed in, but not adherent to, the persistent pales, bright yellow, convex externally, flattened and with a deep central groove on the ventral surface, blunt and hairy at the top.

SEED solitary; *embryo* lenticular on one side, and at the base of farinaceous albumen.

ASPIDIUM FILIX-MAS, *Swartz*. Male Fern.

(Bentley and Trimen's *Medicinal Plants*, vol. iv., plate 300.)

Class III. ACOTYLEDONES.

Sub-class 1. *Acrogenæ or Cormophyta.*

Nat. Ord. Filices.

ROOTS numerous, arising from the rhizome between the frond-bases, black, wiry, filiform, dark brown, long, somewhat branched.

STEM.—No manifest aerial stem, but a rhizome. *Rhizome* more or less subterranean, 2 or more inches in length, and 1 inch or more in diameter, unbranched, bearing at its extremity a tuft of fronds, and entirely covered by the hard, imbricated, dark brown, persistent bases of the fronds which it formerly bore, and which are at first coated with membranous, golden-brown, chaffy scales (*ramenta*).

FRONDS or leaves arising from the extremity of the rhizome, several, circinate in vernation, when expanded 2–3 feet long or more, erect, but curving somewhat outwards, so as to form a spreading plume; *petioles* (*stipes*) long, pale brown, stiff, strongly channeled on their upper surfaces, and densely covered when young with acute, thin, chaffy, pale brown scales and hairs, but these soon fall off from the upper part; *lamina* oblong-ovate or oval, tapering at the apex, bipinnate or pinnate, the pinnæ arising from a stiff, channeled rachis of a pale-brown colour, and with scattered hairs; *pinnæ* alternate, very numerous, overlapping or distinct, subsessile, linear-oblong, truncate at the base, tapering and acute at the apex, diminishing gradually in length up the rachis and ultimately becoming confluent at its end, very deeply divided into numerous sessile, broad-based, broadly-oblong, obtuse segments or separate leaflets (*pinnules*), which are crenate or nearly entire on the margins, finely crenate-serrate at the ends, and smooth on both surfaces.

FRUCTIFICATION arranged in *circular sori*, which are numerous, and placed on the veins at the back of the segments of the pinnæ or pinnules, in a single row on either side of the midrib, usually in the lower half or three-quarters; each *sorus* is covered by a circular-reniform, convex, membranous, smooth *indusium*, attached at the central depression of its cleft, and very persistent, and consisting of numerous very minute, ovoid, blunt, membranous capsules or sporangia, each of which is surrounded for three-fourths or more of its circumference by an elastic ring (*annulus*), which ultimately causes its dehiscence and the escape of its contained spores, which are numerous, exceedingly minute, brown, and ovoid.

CETRARIA ISLANDICA, *Ach.* Iceland Moss.

(Bentley and Trimen's *Medicinal Plants*, vol. iv., plate 302.)

Class III. ACOTYLEDONES.

Sub-class 2. *Thallophyta or Thallogenæ.*

Nat. Ord. Lichenes.

THALLUS erect, 2–4 inches high, flattened, cartilagino-membranous, foliaceous, much and variously divided in an irregular dichotomous manner into lobes; the *lobes* somewhat flattened, variable in breadth, spreading, obtuse or truncate, and fringed at their margins with numerous short, thick, hard prominences. The upper surface is smooth, somewhat crumpled, usually grey or pale olive-brown (but the colour varies much in different plants), and is often blood-red at the base ; the under surface is paler, and irregularly marked with small white depressed spots (*soredia*).

Apothecia (which are but rarely present) placed on the upper surface of the apices of the lobes, large, flat, more or less circular, and of a dark rusty or chestnut-brown colour; *spores* very small, one-celled, 8 in each ascus.

INDEX.